Whole life costing

Whole life costing has been waiting to come of age for many years. A subject that was previously of mainly academic interest is now becoming a key business tool in the procurement and construction of significant projects.

With the advent of public–private partnerships (PPP), and in particular of the private finance initiative (PFI), details of a project's life need to be assessed and tied in to funding and operational plans. Many projects run to millions of pounds and are of high political or social importance, so the implications of the life of materials are crucial. A fundamental requirement of these procurement routes has been that the whole enterprise should be included within the bid, so that a company takes on not only the construction, but also the running and maintenance, of any building.

Additionally, as sustainability has emerged and grown in importance, so has the need for a whole life time-costing approach, driven partly by government insistence. At the heart of sustainability is an understanding of what the specification means for the future of the building and how it will affect the environment. *Whole Life Costing* considers part of this and provides an understanding of how materials may perform and what allowances are needed at the end of their life.

This book sets out the practical issues involved in the selection of materials, their performance, and the issues that need to be taken into account. The emphasis, unlike in other publications, is not to formularise or to package the issues but to leave the reader with a clear understanding and a sensible, practical way of arriving at conclusions in the future.

Peter Caplehorn is Technical Director at Scott Brownrigg. He is responsible for technical standards across the company, currently on projects currently worth more than £2 billion. He writes regularly for *Building* magazine and the RIBA Journal and lectures regularly on construction matters.

Whole life costing

A new approach

Peter Caplehorn

LONDON AND NEW YORK

First published 2012
by Routledge

2 Park Square, Milton Park, Abingdon, Oxon OX14 4RN

Simultaneously published in the USA and Canada
by Routledge
711 Third Avenue, New York, NY 10017

Routledge is an imprint of the Taylor & Francis Group, an informa business

British Library Cataloguing in Publication Data
A catalogue record for this book is available from the British Library

Library of Congress Cataloging in Publication Data
Caplehorn, Peter.
Whole life costing: a new approach / Peter Caplehorn.
p. cm.
Includes index.
1. Building – Costs. 2. Life-cycle costing. I. Title.
TH435.C355 2012
658.20068'1 – dc23
2011041914

ISBN: 978-0-415-43422-5 (hbk)
ISBN: 978-0-415-43423-2 (pbk)
ISBN: 978-0-203-88896-4 (ebk)

Typeset in Garamond
by Taylor & Francis

Printed and bound in Great Britain by
TJ International Ltd, Padstow, Cornwall

Contents

Figures

Chapter 1

Introduction

I have been a qualified chartered architect for over 30 years, and in that time I have designed and overseen the construction of countless projects. In the main, these have been commercially based schemes, ranging from individual or complex apartment blocks, town-centre schemes, offices and schools through to industrial complexes and airports.

Through that experience, I have been involved in detailed work to develop the best solutions for the client and to ensure that the answers provided are regulation-compliant and best value for money.

I have always been interested in the balance between good design and functional excellence. The skills the designer needs to conceive the best solution are considerable – however, in this increasingly complex world, we also need to be able to convince the client and the construction team that the design is valid, practical, good value, and therefore viable.

However, the result of this process often involves a compromise as a result of the many debates and pressures that affect the construction industry today. These may sometimes play out positively, but often negatively, and we all are the poorer for it.

I have always considered the technical and practical aspects of the profession to be the most challenging. 'Form follows function' has been the mantra of many an architect, and is as valid today as ever.

Throughout the whole of my career, I have been concerned over the use or misuse of materials and the squandering of energy. In the early part of the twenty-first century, we seem to have returned to the same issues that I started out with in the 1970s, when Schumacher, Brenda and Robert Vale, Alex Pike and others were making the case for more rational use of resources. We are now revisiting many of these principles under the heading of sustainability and – possibly humanity's single greatest challenge – taming the use and proliferation of carbon (and related gases) and its effects on our planet's climate.

I hope, through this discussion of some of the aspects of whole life values, to develop this debate into a more considered and applied approach that will deliver some tangible and meaningful results.

The need for whole life costing

There has long been a need for greater understanding of materials and resources, and how we use them. As with many issues in the construction industry, this question has been hijacked by third parties, in this case the 'whole life costing lobby'.

In theory, this has produced a raft of information that is supposed to identify the life of the cost of a building, and the cost of the life of that building.

In the real word, however, this is rarely relevant – the normal outcome is to pare costs to the bone, or to justify poor material choices. By the time the results of these decisions have surfaced, those who made them are long gone, possibly retired. We therefore have the construction equivalent of the 'emperor's new clothes'.

This book attempts to clarify this central challenge and to offer some solutions to this dilemma. This is a book rooted firmly in the real world, confronting the real challenges that affect construction professionals on a daily basis. It is intended as a management and project guide that will offer real benefits to projects in the future. It offers:

- an explanation of the workings of the construction industry today
- an account of how circumstances have developed in combination with the practicalities of construction
- some key principles to ensure that sensible analysis can be undertaken to arrive at a real whole-value view of a project.

These factors all have a foundation in financial issues, but are all practical and quantifiable.

So why is life-cycle costing so rooted in money? I suggest that this is largely because the issues involved have been taken over by the financial part of the industry. The client's ear is always open to money matters, and whole life values in themselves are difficult to get on the agenda. We are therefore left with an analysis that is largely removed from the real, practical, everyday world, and will mainly be public relations (see Chapter x). Central to the practicalities of this subject is the characteristic of ageing.

Why is whole life costing important?

Today, so much of what we construct is based on a short-term perspective, and the cost plan is completely dominant. Most project models, especially in the commercial world,

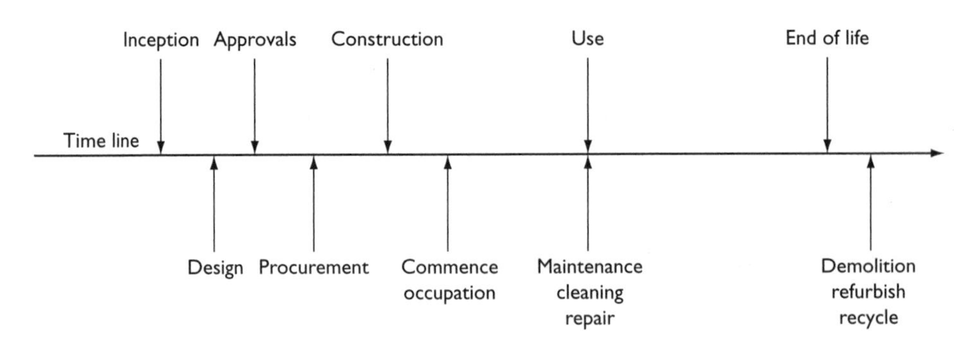

Figure 1.1 The project life cycle

are formulated around the principle of units and the cost of those units. These are later developed into elemental costs, and then into a cost plan.

For many designers, this is unhelpful and disjointed, the relationship of the cost plan and the design being entirely out of step. It is easy to point the finger at the cost consultants here – but the truth is that all members of the design team are usually to blame. Seeing the whole picture, or caring about the requirements of other disciplines, is disappointingly rare. The inevitable consequence is a design that is underdeveloped and a cost plan that is based on too many assumptions – a model that is firmly rooted in short-term profit.

It is for these reasons that the financial detail will not coordinate with the design, any fit between them being very much a matter of chance. This is, of course, both short-sighted and regrettable. The focus is entirely on money. Using finance as the driver to reach more rational conclusions, but not to derive a solution, as most seem to do, is hardly rational. Confused and irrational, these methods just serve to compound the problem.

This book aims to set out a more rational process, away from the financial issues, and to focus on the practical, physical issues that actually establish the whole life cost of buildings and their components. Logically, delivering a real whole life analysis must surely benefit the project and the client, as well as the reputation of the team. In the long term, this must be the only way forward.

Cost is important, of course, but it must be seen in context of the project as a whole, not as a result of – nor the driver for – whole life costing. All forms of analysis to date use a multitude of assumptions to establish a financial statement. This is then used to establish the whole life potential of a particular course of action. How can this possibly be of any real benefit, or in the least way accurate?

It is better to focus on real-world issues to establish the potential, and then to identify whether this is a cost worth paying, and whether it is affordable or even achievable. All too often, the paper principles may not even be achievable, and this cannot be a sensible way to proceed. We need to take action now – if not, we will be forced into reactive measures in future decades.

First, the logic of what is useful and what is not needs to be determined. There is no point in devoting large amounts of resource to analysing a project for it to be so entirely theoretical as to be meaningless. We should be asking at the start: what is the point, where will this benefit the building, or the client, or the end users? Quality of work and maintenance is crucial to all of this, and without a clear understanding of what is required and what can be delivered, there is no point to the exercise. Ensuring that these factors are controlled and undertaken in accordance with the project plan is fundamental. But currently there are few drivers.

Predicting trends in future materials, fashions and commercial pressures is also a complex area. Without some understanding of these, it is difficult to see how any assessment will be of use.

By looking in detail at all the factors involved, a useful model can be produced that allows a range of outcomes to be identified. This can then be used to determine the specification and building operation procedures to deliver the anticipated outcome.

What are whole life cost and whole life value? What benefits do they have? It is important to at least try to estimate the answers, even if flawed.

Any project requires resources. At the beginning, these include the design team and construction processes. For any client requirement, there are a multitude of solutions

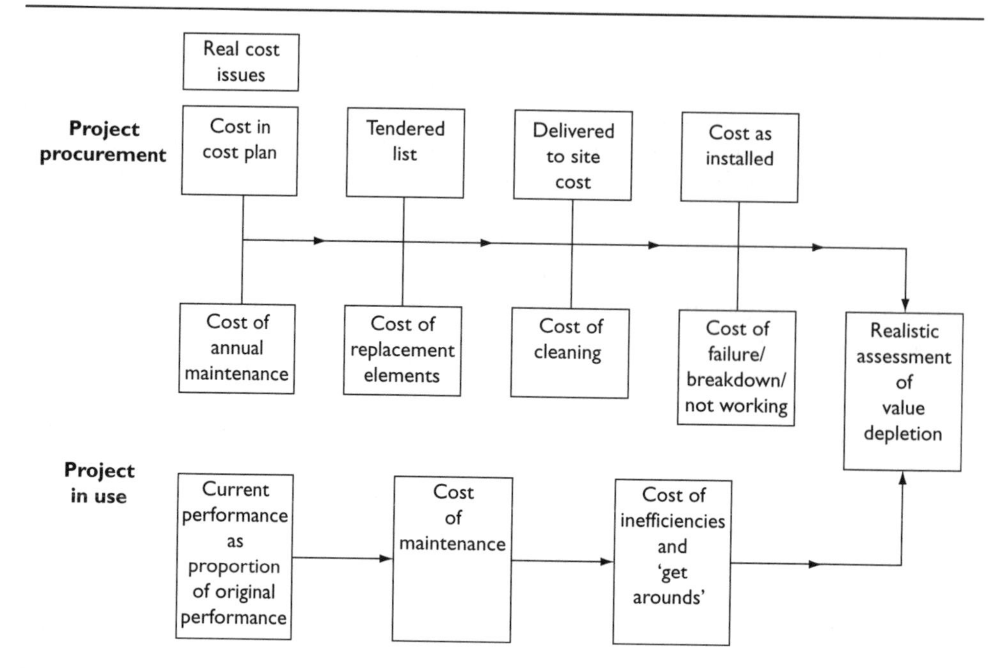

Figure 1.2 Project costs

that will arrive at more or less the same result. However, the details as applied can result in wide variations as to how the building will perform in use, and how long it will last. Additionally, the level of maintenance a building needs to continue to perform will vary, as will the level of maintenance that an owner or occupier carries out. This, too, will have a substantial effect on the life of the building.

There is no single answer to the whole life question. Many variables give rise to a completed project, which will then be subject to many others, all of them affecting the life of the building.

The conundrum does not stop there. How exactly do we establish 'life' and quantify the values of the issues affecting it? This is where fiscal methods alone are inadequate – there is a need to take into account all the various factors to arrive at anything like an accurate answer. For example, it is questionable when a building's life 'ends' – buildings can often be refurbished and reused, which changes the original estimates of whole life value and can change the entire model.

These are difficult issues, certainly, but careful analysis can give some useful results. Linking this with the need to be more careful over the use of materials and energy work done in this area is crucial to the future of appropriate construction. Sustainability and carbon are also part of this story.

It is possible to make some clear statements about what is, and what is not, whole life cost and value, which help to separate the useful from the irrelevant.

Whole life cost is:

- a true assessment of the worth of a building, within limits
- a theoretical judgement using the best information available

- a process to balance the design procurement and use factors
- a process to place cost in perspective as one important element
- a process to be used with care, as results will be a guide only
- useful as part of the analysis to arrive at whole life value.

Whole life cost is not:

- accurate – it should not be relied on to set a business case
- the only measure to assess a project's viability or future
- the only indicator that should be used
- a sensible way to approach future-proofing.

Whole life value is:

- a true reflection of a building's potential heritage
- a realistic appraisal, taking account of all the variables
- a contextual understanding of the future
- a means of understanding the cause and effect of a building's potential future
- a sensible way of identifying a building's future – but must always be viewed in context.

It is also important to identify the meaning of the key concepts involved in this issue.

Life

The time to the first point when the building becomes unviable. This can be due to a number of factors. An asset is unviable when it cannot be used for its prime function and cannot be readily used in any alternative way – when some significant change is needed.

Value

The real worth of a project, taking all the inputs into account. This will vary during the life of a building, but a range of values can be derived, giving an informed view. These can be used to manage the variables during a building's life to achieve the best outcomes.

Benefit

Initially may be in monetary terms, then in terms of resources, quality of performance, or as a net asset, for the purposes of establishing the sustainable options for the building.

Futures

There are a number of possible options, given a building's starting point in use. A variety of paths affecting a building's life will be possible as time progresses, and these may diverge from the mapped direction. The degree of this divergence will be the major influence on achieving the prescribed life value.

Chapter 2

Context of whole life costing

What does 'whole life' mean?

A concept has grown up that, by identifying the whole life cost, rational feedback can inform the choices made at the early design stages. But is this really the case? How does this help, and what does it help to achieve? Any reference to whole life issues may be swamped by much more demanding pressures from a host of other issues. It is therefore not surprising that not much concerted attention is paid to considering either the process or the results of any whole life analysis, and that it has little or no impact on real-world issues in the real world.

Other concerns stemming from this concept are many and varied, not least the need to keep to time, comply with regulations, and ensure that the project is on budget. It is, however, important that the benefits for the client's business are clearly identified, as far as possible, from the very start. The client and the design team should be in the position of understanding the implications of their actions and being aware of the ultimate cost of the project – the whole life cost over the project's total life. But in considering cost, is it not value that should be defined, but true benefit.

When trying to ascertain what happens in the real world, there is little informed information available. It is very difficult to find any evidence that, in real situations, all the analysis and systems produce any tangible benefit. From personal experience of many teams over several decades, I am aware that a considerable amount of effort is employed in discussing this. The usual course of events is that the team confirms the correct path is being chosen and ensures that the given solutions to be built produce a self-congratulatory audit trail that is in fact just a smokescreen – admittedly, a very good smokescreen that fools most of the people, most of the time. Some may disagree with this suggestion, but I have yet to see anything to the contrary in practice. While clients continue to be taken in by this approach, teams will continue to spend time perfecting a story with limited appeal and certainly no tangible outcome.

As the world wakes up to the need to limit resources and to use them wisely, there will undoubtedly be more focus on achieving a good balance between today and tomorrow. Unfortunately, while we continue to base solutions on false premises, the results will not provide the benefits that are undoubtedly needed.

The plain fact is that most of these systems do nothing more than add administration to a project, through an enormous and complex risk assessment to confirm that we are on the right path and that the client need not worry. It is disturbing that so much can be produced for the justification of very little, in many cases nothing at all.

While some will disagree with this assessment, it is made on the basis of 30 years' practice. And it is a point made not with satisfaction, but with great regret. In the future we need to focus on resources, and target construction to be worthy of the resources it consumes. Too much is delivered for too many of the wrong reasons, squandering resources and, in a sense, future options.

Many situations like this have occurred in the construction industry over the years. In particular, this is reminiscent of the early days of the Construction (Design and Management) Regulations 1994, 2007, when risk assessment blossomed like buds in spring, although delivering nothing at all in terms of health and safety benefits.

The fundamental problem in this case, as with many issues in the construction industry, is that theory and practice mix slightly less well than oil and water. It is disappointing that, while considerable effort is poured into joined-up thinking, many key areas such as this remain completely disconnected. If better and more rational thought were applied and used in a constructive manner, we would all benefit considerably.

What goes wrong?

So what are the real issues here, and why does this go wrong? In an ever-more complex world, we need to balance the pressures on a project to ensure the outcome is a true reflection of the project's needs. Not all projects will require long-life characteristics – it is the *fit* of the life issues that is important. In fact, some projects will require a deliberately short life, in which case the design should encourage the reuse of materials, or at least should maximise initial effort to encourage a future life as an entirely different project.

All projects can benefit from whole life analysis, whatever their size, but larger and therefore more complex projects potentially can benefit the most. However, large projects are normally extremely complex and involve timescales of many years that can easily hamper or blur the well meant analysis undertaken if, as is usually the case, there are no checks and balances in this area. In these projects, large teams of people quite often result in considerable churn of personnel, so that those involved at the beginning are long gone by completion, or even in some cases before the start on site. This mechanism devolves responsibility and makes strategic decisions less clear or robust. The net result takes away any real benefits established by the whole-project analysis.

Key industry structures

In considering the established pattern of work, it is helpful to describe the key structures to which the industry works and to generate the main interactions present in any project. There are three core drivers: the conventional project organisation and contract; the team structure; and the contract form. For the majority of projects, these combine to form the backbone of project creation and implementation.

- The Royal Institute of British Architects (**RIBA) Plan of Work** is the established formula for construction projects. It enables the project to be organised around set benchmarks.
- The **project team composition** follows a common pattern of core personnel.
- The **contract form** sets the formula for the roles and responsibilities of the team.

Project organisation

The RIBA Plan of Work consists of several key stages. These are in chronological order, and are used throughout the industry as the basic default standards for breaking down a project into actions and phases.

Stage A Appraisal
Stage B Design brief
Stage C Concept
Stage D Design development
Stage E Technical design
Stage F Production information
Stage G Tender documentation
Stage H Tender action
Stage J Mobilisation
Stage K Construction to practical completion
Stage L Post-practical completion
Stage M Operation and maintenance
Stage Z End of life.

Project team composition

The majority of projects have a team composed of key decision-making and originating people. This is the heart of any project, and the majority of effective actions will be taken by this team.

Client

The client is the originator and the entity in authority, ultimately driving the project and taking the rewards. Clients need to have a clear vision and understanding of the processes and mechanisms involved. Often they are very clear in respect of only some of the issues, and this is where consultants' skill can help and lend value to the project. This is especially true of the one-off client.

Often clients may need assistance and guidance to identify the real benefits from sustainability and whole life value.

Design team

These are the core professionals employed by the client, and sometimes the contractor, to ensure documentation is appropriate. They develop the documentation needed and, in many cases, ensure regulatory compliance and assist the construction team through to completion, occupation and beyond. It is essential that, where required, consultants can employ the best understanding and enable whole life values and principles to be employed, with real long-term benefits.

Contractor

The contractor is employed to procure the components and undertake the project, either directly or indirectly. A large project may use a contractor to manage many smaller companies producing products and components, and then to manage the process on site in constructing the project. Except in a few cases, this is often a process driven by the need for efficiency. There is often little room for considering issues outside the need to complete the build. However, with growing pressures from regulations and performance required of the complete project, there will be a need for greater focus on areas such as whole life value. A single contractor undertaking most of the work will normally be more open to new ideas linked with the benefits of long-term business development to satisfy the client.

Project manager

Project managers are not always used, but commonly control the team and ensure the client's position is represented and the project is directed successfully. They can also help encourage the use of better procedures and the inclusion of whole life values. However, they may be concerned only with the core principles, which may lead to anything regarded as 'unnecessary' being stripped out.

Specialists

Larger, complex projects need specialists at all phases. They may be as diverse as biologists used to understand the site prior to construction, or materials scientists reviewing cladding choices. They can be both a benefit and a possible negative influence by making some parts of the project disproportionately complex.

Contract form

The contract form is fundamental to all operations, from procurement to completion on site. The contract forms establish the relationships between the client, the consultants and the contractor, including methods of payment, the conditions of site operations, and the degree of expertise that will be used.

Traditional

A contract that places the architect or the project manager at the heart of administering the project. The contractor is supplied with exact details and is employed to build only.

Design-and-build

This type of contract is described in detail in Chapter 7. The design team produces proposals that the contractor's team develops and builds.

Management forms

Several forms exist that allow the client more control than a design-and-build, but more engagement with the contractor than a traditional contract.

Several other forms of contract are commonly used, normally responding to specific requirements, engineering, etc.

What, why and how?

Much of the construction industry is formula-driven – it has informal processes and methods of organisation, and informal rules. Many in the industry expect projects to be managed and to progress in line with these processes.

The formal aspects identified above provide the backbone for the procedures, and the details are then progressed in line with these. Over recent years, the design-and-build contract and the package contractor have predominated and have had a major influence on thinking and operations, becoming largely the default way of behaving.

While other forms of contract have their place, design-and-build is very much the favourite choice, and despite evidence that this is now starting to wane after several decades, it is likely to be dominant for several decades to come. The promise of 'on time and on budget' is usually the client's primary wish, so any move away from this form of contract has been thought to be risky.

The principles behind design-and-build are sensible. The designers produce documents that embody the client's requirements, and these are used by the contractor in the bidding process. The contractor produces their own documents based on satisfying the client's requirements. A contract is agreed on the basis that the contractor will build on time and on budget. However, the client relinquishes all authority on the project, the contractor being allowed to use their commercial skills to ensure the project is delivered according to the contract documents.

This is fine in principle, but it has developed an industry that is split around design and delivery. The definitions of performance and 'fit for purpose' have become blurred. The net result is that often the finished product does not reflect the initial requirements in many ways. In most cases this is because specifications are stretched or simplified, with the inherent qualities being misinterpreted or completely misunderstood.

Unfortunately, while some obvious issues may be identified at handover (much to the client's disappointment), many others may remain hidden in the building and emerge only in time, when failures occur or outcomes vary from the story the paperwork spelt out.

Not all projects suffer this problem, but a great many do, and many clients have therefore sought other contract forms in an attempt to achieve both good design delivery and good outcomes. Some may work – but more by luck than judgement.

At the centre of this debate is the reinterpretation of the standards and specification as originally set up by the design team and the client. Clearly sometimes this is an advantage, as the combined skills of the whole team can deliver a really good result. All too often, however, this results in a project where whole life values will be eroded to a very poor level, or consigned to oblivion altogether.

This is personified in the private finance initiative (PFI) form and similar contract forms. These were set up to allow government projects to be procured and run off-balance sheet. The creation of a special company uniquely for the project has nonetheless created a very unwieldy, complex giant that has very little grip on the important detail.

PFI and similar forms introduce a unique problem while attempting to provide a solution for government-funded projects that limits the risk to the electorate. This concept, much touted at the inception, was that the risk to the government was minimal and that performance was built in and off-the-books. The company undertaking a PFI was responsible for the project throughout its complete life. But they could not have been

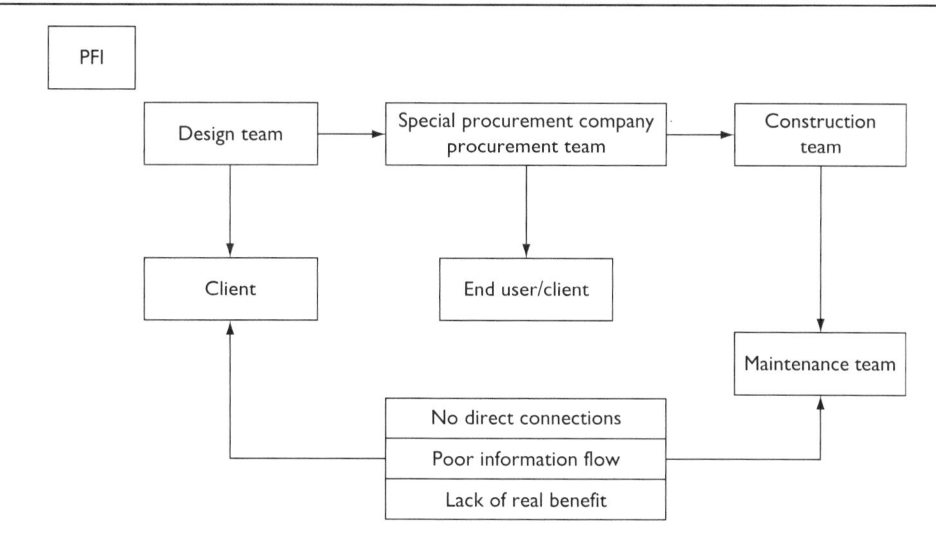

Figure 2.1 PFI disconnection

more wrong. While it is true that some very high-quality projects have been delivered under PFI, there are many that languish due to very poor design and procurement, and construction costs that are cut to the bone. We have yet to see the full outcome of these problems. This is as a result of contracts being sliced up into relatively small sections, limiting liability, with design, procurement, construction and operation being in effect completely separate enterprises. This results in short-term decision-making, and the initial concepts of the project are soon forgotten as the project matures. This makes any logical whole life costing issues completely ill-founded.

At this point it is worth considering the usual pattern for the average project, in order to put these comments into perspective. This also demonstrates the various points that need to be taken into account in any real analysis of the whole project.

The process for most construction projects follows a similar pattern, the accepted path.

Inception – the starting point of any project

The client's team will produce a succession of designs, moving progressively closer to a form of drawings and words that matches the client's requirements and takes account of issues such as regulations, site constraints, and possibly inspiration issues.

This iterative process will need to consider elements of specification quality and therefore the life of materials. This is strongly driven by cost at this stage – these will be estimates, based on expertise within the team.

A good proposal will include a web of philosophical issues that drive the design and are almost invisible to the casual observer.

This is later tested in the market; until then, the prices will be a guide only. At this stage, many projects do attempt a life analysis and even a whole life cost analysis. This is unfortunate in that, even if the system applied is of some merit, the project model will be far from the final design and a very long way from the built solution, making clear projections impossible. Therefore any conclusions at this stage are largely theoretical.

Design development

As the design is developed, specific elements of the design and specification, quotations, and discussions with manufacturers and suppliers will add to the detail within the documentation. It is then possible to establish some clarity in the final project information, but not with any necessary certainty of what the built outcome will be. This is due to the fact that many details of the construction will change in the ensuing months. Products being used will change, specifications will be modified, and build quality that will have a substantial effect is waiting to be delivered. All of this uncertainty will have a significant part to play in the actual, delivered end product and therefore the life of the building and its parts.

During the tendering process, or bidding in the case of PFI, further changes will develop and affect the core project details. Taking detailed concepts into real propositions, with clear materials and products at the centre of the proposals, is a complex process.

Finally, a set of details will be agreed and a construction team appointed. This is the point when procurement starts. But changes do not cease to happen at the end of the procurement process. Further changes will occur when the project reaches site, and will continue until completion. This is because the process is a dynamic one and is never static.

Change occurs for a number of reasons. As the level of detail is improved, facts will come to light that may cause a design change, or a cost increase, or push an item outside the brief. These will all result in changes that affect the balance of the original design. Most will set in motion events that will make the building's future less clear.

Tendering will also bring a level of change. In the best of situations, these will be limited to minor (and in some cases beneficial) amendments. All too often, in the race to be the cheapest and to win the contract, elements critical to a project will be overlooked or left out. A common example is the under-specification of rainwater systems. Often the designer will attempt to specify a system to achieve some degree of future-proofing, as we understand that weather conditions are becoming more onerous. Tenders will often ignore, or just not be asked to specify, the level of performance, knowing that all tenders will price to the lowest common denominator. Only once the price has been agreed does it come to light how woefully poor the system that won the tender may be.

Demonstrably, this point is really difficult to achieve. A reality check is to ask the client how many times a year they expect their building to leak. Often they are quite surprised, and begin to delve into the point behind the question. When the actual issues are debated, it is clear the accuracy of the tendering process often lets the required standard escape. A presumption exists that the building will not leak; however, the tender documents do not prevent it from doing so.

Detailed contractor issues, especially in design-and-build and management forms of contract, continue to distance the designer from the actual product. This always leads to issues over quality and fitness for purpose. We continue to hold the perception that designers always over-egg the design, and that, conversely, contractors always cut quality to the bone.

Manufacture of significant or complex items can also lead to changes that may remain hidden well past the building's completion, and surface only many years down the road, such as a mix of materials or the quality of a seal. Examples may include changes in fixtures and fittings, where dissimilar metals lead to corrosion and eventual failure. When construction starts, site operations also generate a mix of changes, especially on site, which is also a problem. To all practical purposes workmanship is unsupervised in the current industry, and the level of training is at an all-time low. The ability of a pressured, poorly trained, under-skilled workforce to deliver the required quality is optimistic in the extreme, and a constant source of problems. We see the use of operatives from Europe as a benefit in that many are skilled and can easily deliver the quality needed.

Quality assurance is a good system if used correctly; however, very often it is merely lip service and even auditing is not astute enough to ensure successful delivery.

Verification is often used to establish if the required quality has been achieved. Again, there is little in this process to give confidence. If we look at the poor level of compliance with building regulations, we can see clear evidence of the gulf between theory and practice. For very solid reasons, we need better compliance and reliable quality.

Designers need to use references and standards to ensure the detail is comprehensively transmitted to the construction team. They do this using standards and quality appraisals that describe the relevant requirements.

- There are several good references and guides already in place. But all guides and references should come with a health warning: they need always to be read in context, and with the benefit of other references, before being relied upon.
- Several commercial web-based systems and databases are available.
- The Building Research Establishment (BRE) and other world-renowned research establishments publish a number of useful guides.
- Several British and international standards are also of merit.

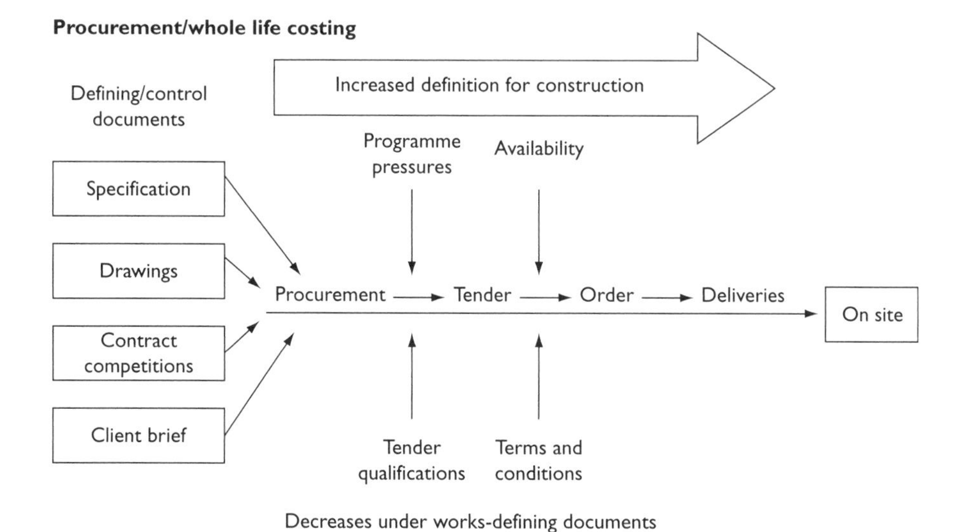

Figure 2.2 Procurement

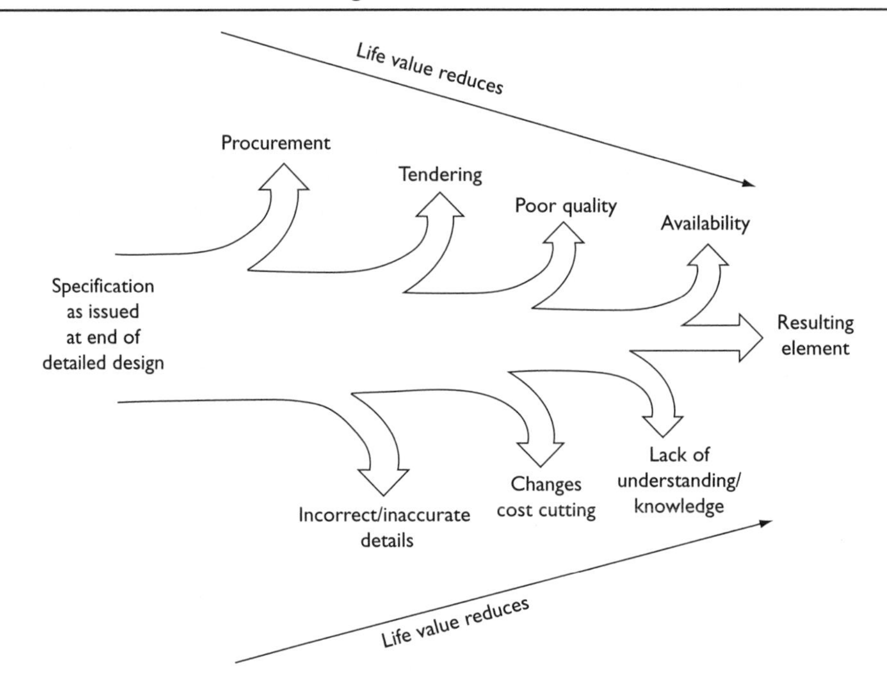

Figure 2.3 Dilution of specification

However, we come back to the central question: how do these standards work in practice? What real benefits do they bring when confronted by the everyday pressures and complexities inherent within any project? Many would say that they need to be more straightforward, to avoid waste and avoid deflection, to achieve the desired aims.

Value judgements are undertaken constantly as part of a designer's daily routine. These are a multitude of whole life assessments, although rarely acknowledged as such. This is the everyday situation. However, all too often they are overruled by value engineering, time pressures, contractual restraints, and poor decisions made entirely on a cost basis without looking at the wider implications. This results in the documentation used to drive procurement as being the minimum for the brief. Often in today's industry it is deliberately left undecided, so that the subcontractor or specialist can finalise the details. The industry has seen the growth of design-and-build procedures as the common process.

If a project is still being designed as it hits site, and many are, this is obviously not a good situation for any whole life perspective. Issues may be resolved well into the build programme, with components coming to site and site works being resolved on a very short timeframe. As the works continue, this process can become difficult as there is a need to resolve issues ahead of the project programme. This means that answers centre on the expedient, not necessarily on the best, either in the short or the long term. Consequently, we have many examples of materials used inappropriately, or a fix-it-for-now approach being used. This does not leave any room for consideration of basic maintenance, let alone the principle of any whole life issues.

At handover there is always a focus on completion and occupation. There is real pressure to get the building up and running. Rarely is any thought given to maintenance programmes except when defects are discovered. These bring into focus issues to be resolved and the impact they will have on the ongoing performance. This is currently focused around performance in the short term; very little evidence of long-term considerations is available. This is one of the major inhibitors to progression of real whole life considerations. However, the growth of building information modelling (B. I. M.) will hopefully change this position. Change takes a long time in the construction industry, and this sort of embedded change takes the longest. The need is there, the logic is there, but until some key drivers are established there will be no progress.

The fact is that many of the guides and initiatives regarding whole life costing have been around for decades, but little has changed. There are some examples of good attempts, but the underlying fact is that, while they all start with the best of intentions, there is very little evidence to suggest there is any real benefit to the project or the client. This is a good idea with tangible benefits, but there are many reasons why it is still theory. I have set out in Chapter 15 how I believe this could be reversed and a framework could be put in place to achieve better – not perfect, but better – results.

In conclusion, there is a need to be more practical, focused and objective-driven. There is a need to stop avoiding the obvious and address it – until this is done, no system tracking the future of a complete project will succeed. Whole life costings are put in place only for them to be ignored after a very short time – rarely do they survive beyond two years into the project. This needs to be accepted, and a move away from the prescriptive approach is needed.

A series of gateways is possibly more robust, driven by a formal arrangement on the project that will stand the test of time. A link to fiscal gateways and licences is also a possibility.

In Chapter 22, I touch on the use of energy performance declarations or suitability, and pressure to take account of carbon life costs, with gateways as the possible future. There is a need for buildings not only to perform when new, but also to continue at that level, or a respectable level in proportion to the specification, for some considerable time. In this way it is possible to see that whole life values can assume an importance that they currently do not attain.

These requirements are developing a real importance that very shortly will be encapsulated in standards and legislation. It remains to be seen if whole life could or should be regulated; however, there is an obvious issue that when resources become limited, legislation may be the only lever that can make outcomes certain.

Chapter 3

Sustainability, energy and waste

Sustainability is the well considered use of materials and resources today for the benefit of the generations of tomorrow. No thinking about future issues can be complete without fully considering the implications of sustainable options. In particular, whole life assessment is crucial to the foundations of sustainability. At this stage of the early part of the twenty-first century, sustainability is one of the top-line subjects: not only does it involve the essential issues of carbon and climate change, it also focuses our attention on the use of raw materials, water and land. It is absolutely vital that when we expend resources, they must be employed well and to the maximum overall benefit. The role of whole life assessment value and cost is central to this task and must taken seriously.

It is essential that sustainable issues are part of the core agenda within any whole life value or costing appraisal. Sustainable thinking and practical application are the most challenging issues in today's construction industry. It is vital that methods and principles are adopted that embody a sustainable approach. Assessment of whole life value is an important aspect of this significant area. Understanding how a building will perform, or at least obtaining a reasonable assessment, is invaluable in this area. From every perspective, to maintain the status quo or ignore the issues that we are creating for future generations is plainly absurd.

Sustainability and whole life value are obviously interconnected, the latter being a fundamental affirmation of the former. Ensuring that the life of a building is appropriate is at the heart of sustainable thinking. There is little point in devoting resources to the construction of any project unless they are of benefit. As resources become limited, there is an increasing need to understand that they can be put to good use over a relatively long timeframe. Justification of the application of resources must be in proportion to the prescribed life. However, we often see constructions that use the lowest performance values and expect to achieve long life values when they plainly will not do so.

Growing political pressure to ensure sustainable options are adopted makes the need to regulate increasingly certain. There is every reason to believe that shortly this will become part of legal requirements under the development of the Secure and Sustainable Buildings Act 2004 – this Act has achieved Royal assent but no provisions have yet been put in place. Until and unless this takes place, it is uncertain whether there is enough momentum to drive involvement in the subject to the level needed.

Many buildings have a notional life that will be established by a project specification reflecting the client's requirements. Particularly in the case of commercial buildings, the need to be accurate and accountable is essential to the viability of a project. The client and funders need to have confidence that the design life has been established realistically. However, little or no account is currently taken of the real cost of this period – which includes accurate demonstration of the potential issues and the various outcomes that will have a life-modification effect and cost. Logically, there is never going to be one answer to this question, and single answers should be regarded with suspicion. A range of answers is, however, both realistic and potentially very useful.

In attempting to establish an answer to the fundamental question of what period of life is required, the common answer is 'as long as possible in proportion to the financial cost'. That is to say we should specify and build to the best level of quality and robustness that the project can stand. This is because to do otherwise is wasteful in the extreme and highly unsustainable. Often not enough time is spent on this issue.

Despite this inescapable logic, this is the situation for the majority of buildings currently. Any sustainable threads are a consequence of other drivers or are purely fortuitous. Only in very recent times have there been any real attempts to ensure a building's life is as long as possible, therefore maximising the resources and materials used in its creation.

Life costing and the establishment of future performance is of considerable importance to sustainability – it is potentially the key driver. It is not unreasonable to assume that in the near future there will be a requirement to demonstrate as small a carbon footprint as possible, that will include a clear understanding of every component's life. The importance of ensuring efficiency with regard to every element of a project is clear – but how is it to be achieved, in the face of existing pressures and inaccuracies?

As the pressure for greater efficiency increases, considering every construction as a resource for the future is becoming the norm. Analysis of the resource's potential is likely to become established as a key driver, in the same way that carbon footprints have currently taken over from simple energy-use calculations.

The need to be more sophisticated is pressing and essential. We cannot afford to squander resources, effort or time in the pursuit of values that do not really deliver results. Many current approaches to this problem are very crude, either taking too simplistic an approach, or using a model that is far too specialised and complex for everyday application.

One related area that can be seen as analogous to this conundrum is the analysis of energy in buildings. The majority of energy considerations, until relatively recently, concern simple heat-loss and heat-gain, largely through the mechanism of U values (the U value measures the heat flow through a building element under steady-state conditions that limit its accuracy in comparison with real-world energy flow, which is always changing). Ensuring that energy and waste are considered and managed intelligently is essential.

Energy use in production, embedded energy, and the energy in use as the building is occupied are obviously of importance. The longer lived the building and the poorer its performance, the greater the cost to the environment. A building should therefore be as efficient as it can be.

The life of a building's components is exactly the same, but less quantifiable.

Sustainable options

Relevant sustainable issues:

- minimum impact on the surroundings and our environment
- maximum use and benefit of materials

- maximum and most efficient use of energy
- reuse of materials, products and components
- minimum maintenance required
- minimum repair required.

The consumption of materials that offer no contribution to the completed project is of great concern for sustainability. Under the general heading of waste, measures are now in place to raise the issue and try to ensure waste is reduced and, in time, reduced to an absolute minimum. The imposition of the Site Waste Management Plans Regulations 2008, governing the control of generated waste, demonstrates the government's view and the seriousness that is attached to this problem.

Any building project will generate some waste. Putting a whole life value on this element is important – waste is part of the total resources needed to generate the project. Ensuring these are limited makes the overall project more efficient.

Primary and secondary energy, to produce, run and maintain a building project, are difficult to separate out or analyse in any useful detail with current levels of data and understanding. However, the primary energy used to create materials and install them on site can be analysed against the energy consumed during use. In the case of aluminium, for example, its high primary energy in manufacture is balanced against its long life and very successful reuse.

Once created, the material can go on through many cycles of use, and repay the energy input required in its creation. Many questions still remain, however, in the definitions of use and reuse in this simple example. We are likely to see clarity coming from the carbon debate to identify the notional start point in any single cycle.

Other materials of similar overall balanced characteristics should also be considered. For some materials, embedded energy is worth the investment, whereas in others it is not, and cannot be seen as any more than a poor choice of specification.

Similarly, some materials have a balanced effect in use, so while the initial energy input is large, they contribute well (and possibly uniquely) to the overall performance throughout a building's life. For example, cement has very poor characteristics and requires large amounts of energy to manufacture. However, used as part of a concrete frame, it is very beneficial. The frame itself will have a long life, the mass in the floor slabs can be used for cooling, and the material may be recycled as hardcore when the building is eventually demolished.

Rising sea levels, limited food supplies, water availability, and waste creation and disposal are becoming real, crucial issues. Over the next few decades they will become vital, affecting the survival of many countries. Combined with the effects of ever more aggressive weather conditions, any prediction of future effects and patterns starts to appear limited. Due to the overarching nature of these issues, they are far more influential than they at first appear in and around the built environment. It is most important to keep this in mind in the assessment of future value, and this makes the use of a parameter-based rather than a quantitative assessment much more realistic. It is essential that any future models take account of all these issues and make the choices clear to the client in a straightforward manner.

'Built to last' seems an alien concept, but was once the watchword of the building industry. More often than not, it is now an unfortunate fact that buildings are built to fail, or at least have the mark of failure all over them, from handover.

Today, investments are made with the clear objective of providing savings down the line – but how often is this actually rolled out? The sustainable option delivers exactly that, and is coming to be seen as a major driver towards better whole life evaluation and application.

Sustainable considerations for life value that need to be built in to the original brief include:

- longest possible life within specification
- widest possible uses
- minimum safe maintenance and cleaning
- minimum environmental impact to procure and construct.

The skills needed to deliver knowledge in this area are generally few and far between. Very many sustainable options are not thought through or achievable. Many falsely sold options are adopted within projects, which only serve to compound the issues faced by the industry. There is a need first to understand the basics and then to be able to translate them into clear, achievable direction.

Asset management is part of this equation: the building in use must be managed correctly, which requires some degree of understanding of the building. Checks on how it is performing and references to its life prediction cannot be made successfully unless verified on a regular basis.

Operational efficiency is generally a very poorly understood science. Many buildings, while they have been carefully crafted in conception, and perhaps delivery, will be largely ignored in these terms. The building's performance is largely off the radar as an issue until something breaks or seriously goes wrong.

The need to provide energy performance certificates, along with greater emphasis on sustainability issues, combined with higher energy prices, will create change.

Suitability issues

- Life versus embedded energy.
- Transport of higher quality (long-life) materials from source to site.
- Resources required for maintenance versus shorter life.
- Reuse versus recycle.
- Environmental impact: high-impact, long-life versus low-impact, short-life.

The answers are not necessarily simple – balancing the criteria from one point of view with those from another is always difficult.

Currently, there is not enough data to make these decisions easy or obvious, and many can be counter-intuitive. The establishment of suitability credentials at the start is complex and open to considerable debate; life assessment adds another level of complexity that many are not yet ready for.

Chapter 4

The rational analysis

Essentially, this is an exercise in establishing exactly what we know and limiting the assessment, making sure that the conclusions are factual and there is the will and the ability to make a project work in the future, in some cases for a very long time.

What do we need to know?

To understand the issues that confront any whole life issues, we need to have a clear understanding of the background logic. The first exercise is to consider how to analyse and break down decision-making into a sensible framework.

We have several types of knowledge:

- things we know to be true
- things we don't know
- 'unknown unknowns' – things we don't know that we don't know.

We can plan for all the known areas, and to some extent the known unknowns, but the unknown issues that may crop up are, by definition, impossible to identify and yet may have a significant impact on any whole life analysis.

For this reason, the whole area where we do have some knowledge must be allowed for and identified in building up the analysis. In short, we are attempting to ensure everything we know is allowed for, in the clear understanding that there will be as yet unknown issues that cannot be allowed for, and their effect will need to be minimised.

Why do we need to know?

Understanding the context of our knowledge is equally important. Much of construction is based in the context of a series of interrelated issues. Knowledge of one without the other makes any understanding flawed and ultimately pointless.

It is for this reason that much information, guidance and analysis is not worth noting. Countless propositions that are either not fully researched, or do not cover the whole picture, are issued and regarded with high esteem, whereas they should be interrogated with vigour or ignored. An informed point of view must be obtained at or near the coalface. Equally, an informed perspective will understand and take account of the fact that it is unlikely to represent comprehensive knowledge of all outcomes. Furthermore, the variance due to lack of knowledge is likely to increase with time.

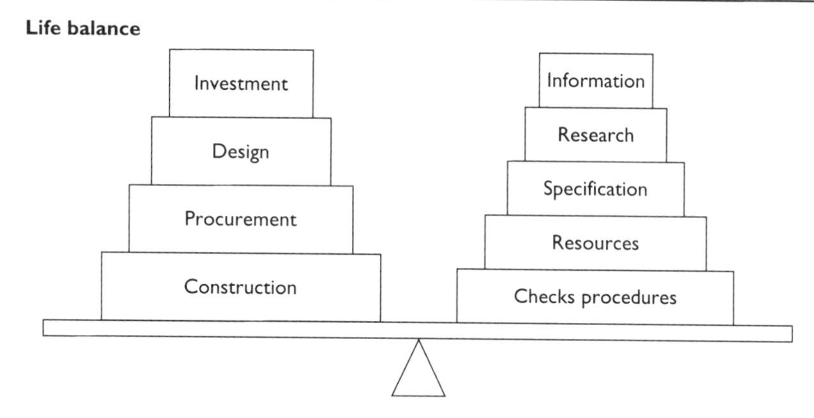

Figure 4.1 Value/cost balance

Value choices are made throughout the process, from inception to completion, with regard to the components, materials and assembly of builds as a whole. These choices are considered by the designers and then reviewed by others in the wider team. On many occasions, the judgements placed on certain elements are disproportionate. A clear guide to these issues is needed.

The various influences in play at any particular time need to be organised and considered; often opinionated decisions are made based on very little real information. This, then, is used to turn a decision-making process into one of personal judgement and/or the verification of a direction that the individual or company wants to pursue to their own benefit.

All too often, these situations occur and are masked by a plethora of information and so-called analysis to establish the required result. It can be seen that by the time reality catches up, the original protagonists of such woefully misplaced decisions are long gone.

The net result is that, even with the best model of understanding, projecting forward 10, 20 or even 30 years into the future, there is increasing inaccuracy and consequently projections should be relied on less and less. This may seem obvious, but too much emphasis is placed in complete confidence that the future of a project is clear and defined, and will run on the prescribed course. Rarely does this happen, and any rational analysis must take this into account.

However, this should not stop us from attempting to define the future and getting as close as possible to where we think the future lies. It is therefore best to consider a range of possible outcomes. These may be ranked and can be considered as a diverging set of paths increasingly spreading away from the desired route with time.

The ability to grasp the limits of our understanding and what can reasonably be expected of any given analysis is crucial. It is essential that a degree of realism is maintained, and can be tested from several different perspectives. Taking this into account within the construction process, it is possible to establish which are the appropriate issues to focus on, and which to ignore.

Design is a complex and highly skilled process, often underrated by those lacking full comprehension of the detailed iterations it requires. The first task is the synthesis and comprehension of the client's demands. It may often be that the client does not know what they actually want, or perhaps careful consideration will show that the brief they have established is wide of the mark. The skilled designer can address this through a series of conversations teasing out the client's real requirements. In some cases, this may result in the direction of the project changing, or in the project itself being questioned and perhaps shifted or even cancelled. This is perfectly sensible, but does demonstrate that design is not a linear process.

There is a need for the designer and their team to bring a clarity to the table that others may not have. The rare ability to see beyond the cold facts to absorb the client's requirements, including some issues that they do not recognise themselves, and to set the process in motion, is sadly all too scarce.

After the initial briefing, the care taken in developing images and setting down the possibilities is the start of an iteration that will continue, to a greater or lesser degree, up to completion and sometimes beyond.

The design process continues with the development of detail at all levels, creating ever-increasing levels of detailed information and coordination. Within these are a complex series of relationships that need to be understood – but rarely are. This can be broken down as follows.

- Design:
  - identify the performance – size, shape, quality.
- Selection:
  - identify possible sources – materials, quality, availability.
- Manufacture:
  - identify a source that combines best design with best selection.
- Assembly:
  - identify the optimum path for combining the components.
- On site construction:
  - complete the process.

In each of these steps, care and understanding must be applied to achieve a balance between over- and under-specification. The challenge is to ensure the key points of the design are translated through the rest of the process and are reflected on site in the complete building.

Life value may be defined at each of the various stages, and knowledge is critical to this understanding.

- The brief:
  - must identify the design life, the maintenance periods, and the requirements at the end of the design life.
- The uses:
  - should be quantified as they have a significant bearing on the level of durability required.

- Possible materials selections:
    - for any given use and life, there will be a range of possible material component choices, which need to be identified and ranked in terms of appropriateness; normally the range will include a spread of cost and performance.
- Possible assembly principles:
    - there will also be a range of assembly options, which may be heavily influenced by the procurement and construction teams. Care must be taken in this area to ensure the attributes needed are maintained.
- Procurement routes:
    - ensuring the procurement route does not affect the life values is possibly the biggest challenge. The team should ensure these are kept high on the agenda.
- Contract forms:
    - can also have a significant effect – the common use of design-and-build can mean substantial changes to the details of the design and specifications if the life value is to be preserved as design, and this must be ring-fenced through the contract. This can be quite difficult to do, and for many projects, using a contract form where the client is not in authority will probably cause problems.
- Chances for desired outcomes:
    - given the number of issues that can deflect the original choices, ranking the range of options and making the client aware is an important task. This should at least give a perspective of possible situations that may push the project off target.
- Chances for failure:
    - if carefully organised reviews are undertaken, the chances of absolute failure are extremely small. That is, for a significant component or components to fail completely and prematurely is highly unlikely.
- Need to correct errors:
    - equally, when changes occur they need to be evaluated and corrected if necessary. Just letting them 'roll by' may have very significant effects later in the project's life.
- Politics, costs and timing:
    - are still the main drivers for any project; clearly any life-value equation will take second place. However, if the brief is serious about real whole life issues, it must be balanced against these fundamentals – unfortunately, very often it is not.
- Knowledge allows:
    - predictability – how good and solid are your processes?
    - will they stand up to the expected use and wear-and-tear?

Only by ensuring that the routes through to completion are reasonably within the band of expected criteria will a well defined outcome result. A common broad-brush view would be:

Finishes	Substrate	Substructure
1–3 years	5–15 years	20–40 years

What are life and life values?

One of the aims of this book is to provide a practical guide to the establishment of life value and the problems that can affect it. We begin with a definition of exactly what life of a building actually means, and how it is established.

With any project, we first have the design and procurement process, which is theoretical, followed by the actual construction and use of the building.

- Theoretical (life as set by the project's parameters):
  - period to first significant maintenance/refurbishment
  - period to first significant failure
  - period established by industry/good practice
  - period required by the client.
- Actual (life as achieved by a material product or component in use):
  - period as defined by the manufacturer
  - period as defined by independent testing
  - period as established by accelerated ageing

then:

- period to first significant maintenance
- period to first significant failure
- period to first unviable use.

These criteria are normally spelt out in the specification. This is derived from the brief and informed by the design.

General and detailed criteria are needed and should be identified. These need to cover the general principles as identified above, and then lead to specific quality statements on workmanship and site-control issues. Correctly done, this will ensure the life values are protected.

If the process is undertaken correctly, there is no reason why the life cannot be exceeded thus increasing the life value.

- Establish specification (purpose, life and quality, for example as listed by the Royal Institute of Chartered Surveyors):
  - physical
  - economic
  - functional
  - environmental

- technological
- social
- legal
- practical
- materials
- components
- assembly
- product
- transport to site
- erection
- completion
- maintenance
- cleaning
- repair
- demolition.

However, it is also worth considering some of the practical issues that come into play on projects.

- Some methods to use on real projects:
  - analyse the brief and client requirements from a life aspect
  - consider the construction type from a life-value aspect
  - consider the materials in use and how they relate to each other
  - consider the likely maintenance regime
  - produce a worst-case/best-case scenario
  - identify weak points – can they be improved? If not, plan for them to fail early
  - changes during the procurement and construction phase should be checked and supported or resisted and explained clearly to the client
  - if the answers are really not clear from the analysis – start again.

Chapter 5

Value of whole life assessment

Despite some shortcomings, there is real benefit in the assessment of a design in terms of its potential life – it is how we establish this that matters, and once established, what is that value? Value in real net asset terms – value financially and value to society as a resource.

The key areas affecting value are:

- affirmation of the life as designed
- quality matching design at completion
- maintenance required
- potential risk areas
- potential at life end
- accuracy of the business model.

Known knowns

At the heart of life assessment is the filtering of knowns from unknowns – in other words, replacing uncertainty with certainty. This is easy to say, but far more difficult to achieve. Why we need to know, and what we really know (as opposed to what we think we know) is important, but difficult to establish. This is crucial to establishing the real value of a project.

How can the designer assess real value from the beginning? The common starting point has to be experience, hence the repeatability of a known situation; but, paradoxically, the very essence of good design is to go beyond experience and try new and potentially more difficult solutions. This will test the designer's ability to understand the possible outcomes in terms of both life and value.

Balancing this issue is the value created by original or aspirational thought. New buildings that have a cutting-edge approach can be very successful such that the value they create far outweighs the risk in attempting something untried. With this potential comes the problem that the new ideas may not succeed or find favour, or may become unfashionable all too quickly. By analysing the issues, an assessment can be achieved to maximise the pluses and minimise any downsides.

Market pressures

It is a very rare project that has no time constraints. These manifest themselves in a number of ways, some real, some imagined. The fear of time, or the lack of it, drives many decisions and can be the cause of many project woes.

The benefit of experience and the foresight of good, clear-sighted planning can go some way to offset these issues. However, the pressure remains, and can have a significant impact on the whole life viability of any project. The best intentions can be destroyed when particular materials or products are not available in the timeframe required. Inevitably, this means that any substitution to fit the programme will have bad consequences for a project's life values. There is no easy answer to this, except to ensure that time and availability are checked out early in the design, and that supply-chain demands are well understood and organised.

The better the engagement from supply chains, the easier this process becomes. All too often these relationships are too fragmented and financially constrained to ensure well coordinated and cooperative working. Supply-chain engagement is a very important contributor to whole life value. Without the delivery of the appropriate materials and components to site at the right time and to fit the programme, the most carefully determined characteristics will be missed.

This is also important for the building in construction and use. The supply chain delivers a high-quality, well trained workforce to ensure the building is correctly assembled, finished, commissioned and maintained.

Time issues for a project are ever-present and critical. It is very easy to lose time and intensely difficult to make it up. Projects that run behind almost never get that time back and recover.

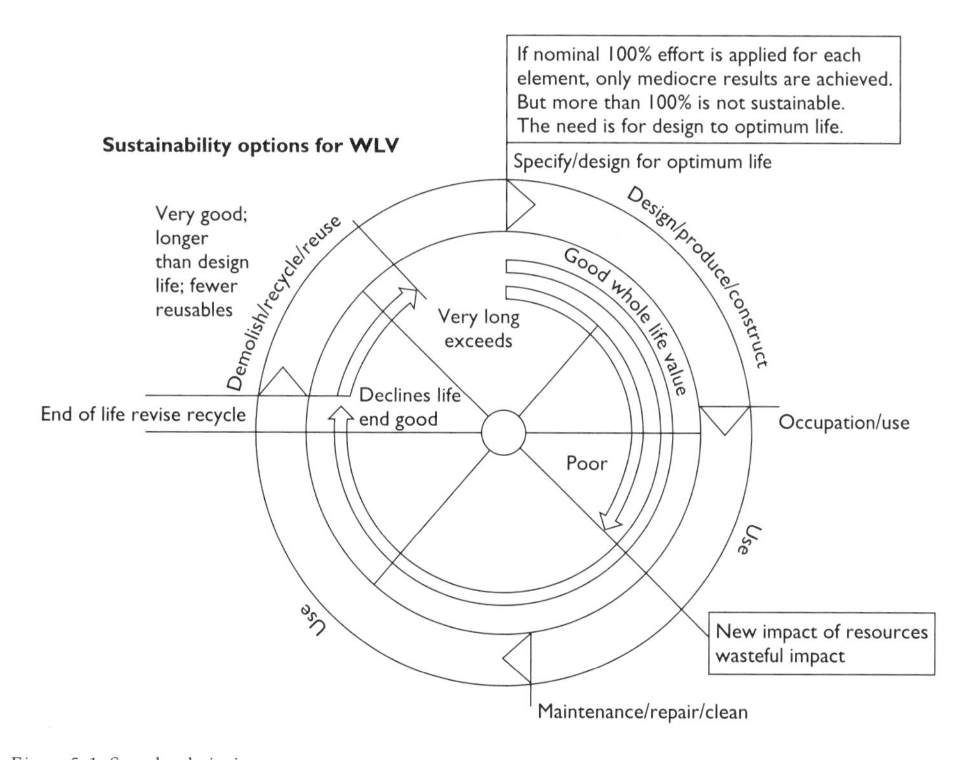

Figure 5.1 Supply-chain issues

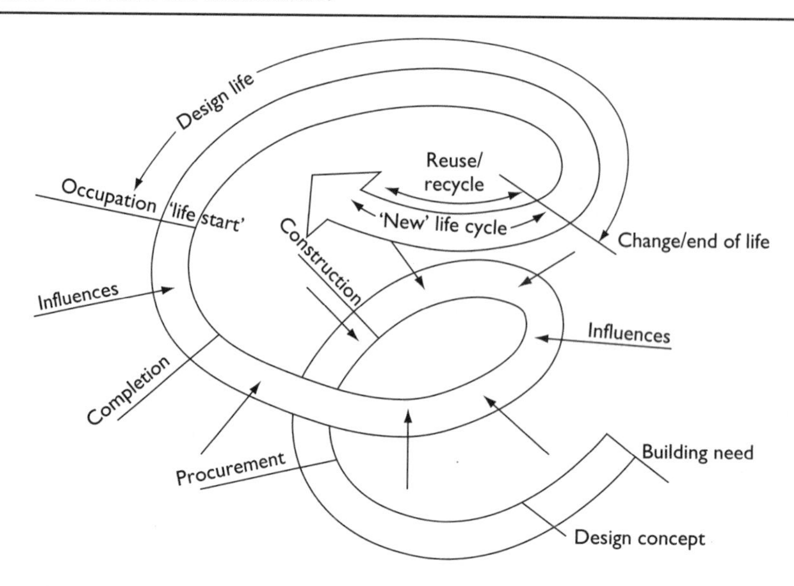

Figure 5.2 Influences by third parties

This pressure manifests itself as:

- decisions driven by time
- short-term conclusions taking priority over long-term ones
- future problems ignored or dismissed
- quality ignored for expediency
- ensuring delivery availability over all else
- rational responses limited.

A common consequence is that time pressures get the better of the project team. It is of great disappointment to consider that many of the issues occurring during project planning are a result of lack of knowledge or fear of failure. Many projects are made the poorer due to these pressures, and what suffers unseen is the effective life of the materials and the building's systems. This is because these are the easiest issues on which to 'cheat'. The consequences of these actions will not been seen for many years, sometimes decades, at which point the original principles or actions have been long forgotten.

At the heart of this issue is the sourcing and ordering of the specified materials and products for the project. It is clear that the decision-making process arriving at the designed item is complex. However, once the programme of construction has started, it is often impossible for the team to stick to these decisions. The need for substitutions is realised by the truckload, driven primarily by time pressures. This will almost always be negative, and will rarely result in anything other than a reduction in quality and performance. This inevitably results in a reduction in the effective life of the items in question and their overall value.

Most projects are delivered on time and on budget, but are only superficially compliant. A look behind the scenes will inevitably show signs of issues buried during construction, which will come out of the woodwork in years to come. The team may often feel this is a small price to pay for delivery of the project on time. There can be little criticism if the building is completed within budget and opens on time.

It is not tenable that we will ever see a situation where failure to reach the short-, medium- and even long-term goals for a building's performance will result in the original team, or those who represent them, being questioned over their decisions. While this may seem a little heavy-handed, it is important, as many opportunities to build efficiently, and for the future, are frittered away due to lack of rigour. While this continues, the long-term life of our building environment is in question, and no-one can claim we are engaging with the sustainable agenda to any real degree.

Underlying this issue, driven by time and fuelled by the contract structure, are nagging doubts of uncertainty. If there could be a certainty that the performance would match the specification, then there would be little to fear.

Behind delivery of the specification is an atmosphere pressured by time and money. Control of two issues that determine component availability is crucial: the first is the choice and detail of specification; the second is the possible changes that occur in the procurement chain.

The designer may have correctly decided on the material element or product (and this is by no means certain), and they may also have selected this bearing in mind the life and recourses balance that is the essence of getting the whole life cost correct. But this is pointless unless availability to suit the project is correct. Ensuring that the desired item can be obtained, and in due time, is of core importance. If this is not achievable, there is little point. Often choices can be made, and even confirmed over a long period, only to be wrecked at the last minute due to the fundamental lack of availability. This can be caused by unrealistic demands from the specifiers, unsuccessful delivery by the company, perhaps something in the terms and conditions, or simply the fine grain of price issues. Whatever the reasons, these problems are present at a very difficult point for the project, and even simple elements can cause significant problems.

Key issues here are confirming that:

- the project needs whole life value assessments
- the whole team is agreed on how to do this and what is involved
- the output will have real meaning and can be trusted
- the implications for use and maintenance of the building are understood
- key choices will be made with the right information.

Procurement – friend or foe?

Procurement is also a period when availability can have a significant effect because it is used as a lever. Whatever the rights and wrongs of this situation, the plain facts are that rarely does a project come to completion without this being responsible for a change that will have a significant effect on the building's life.

Pressure on most projects is considerable – pressure of time and money – and it is this that is the cause of most change. This occurs because attention to detail and poor communications remain at the heart of the UK construction industry. This is disappointing in that huge resources have gone into attempting to ensure the reverse. Nonetheless, this endemic problem, almost from the moment the contract is signed, starts to eat up money and erode the programme. It is little wonder, then, that changes start to be built in from the outset. Changes to cheapen; changes to make simpler; and changes that have to happen because the original choice is simply not available.

Chapter 6

How does this play out in real projects?

To understand this problem, and to illustrate the issues, this chapter takes an in-depth look at an actual scenario and the issues that surround the decision-making process. Taken from real project experience, I hope to show the everyday pitfalls and complexities that, despite good intentions, can result in completely overwhelming whole life issues.

A common element in the design specification and construction of a commercial building is the roof. This is an area that normally has a long life requirement and needs to be handled with great care to ensure reliability. The interaction of waterproofing with insulation, and ensuring resistance to sometimes quite extreme conditions, are crucial elements. Again, often this is seen as an area in which to cut corners and reduce the design requirements.

The order of activities is straightforward:

- brief-setting
- establishment of the specification
- detailed design
- procurement
- tendering
- ordering and availability
- site establishment
- work on site – quality control
- completion
- maintenance and use
- recycling and reuse.

Brief-setting

Certain characteristics are fundamental requirements for success. Obviously the building must be waterproof, robust and durable. Most offices are designed for primary elements, such as the roof, to have a design life of at least 25 years.

- The roof needs to provide a good thermal performance in relation to both heat and cold.
- It must be easy to clean and maintain.
- It must be able to act as an energy collector, possibly even a rain collector.
- It must not transmit excessive noise from weather or external sources.
- It must contribute to the building's 'green credentials'.
- It must be safe to build, clean and maintain.

Establishing the appropriate specification

- Sets the right direction and detail.
- Ensures that the correct life is built into the equation.
- Gives a balance between the materials and their performance.
- Controls the quality of workmanship and physical criteria.
- Ensures that appropriate checks and balances are used to establish performance.
- Establishes the benchmark for the requirements.
- Supplements the drawings and other documents to give a complete picture.

Nine times out of ten, problems can be traced to a change in the specification amplified by poor workmanship and then followed by insignificant maintenance.

Failures in the early years of occupation demonstrate a lack of resilience in a completed building and the potential life-shortening that has already been put in place.

Having achieved the design life, there is the potential to enhance a building's usable life further, but this may not be possible or desirable if other factors concerning the viability of the building prevail.

Inevitably, it is commercial pressure or fashion – as discussed in more detail in Chapter 13 – that can curtail a building's life.

The detailed design

A carefully balanced specification is needed that considers all the elements: structure, insulation, vapour control, membrane, finish and drainage. Ensuring that all of these work together and do not have any long- or short-term conflicts is part of the designer's responsibilities. A designer will often choose products from the same manufacturer to ensure compatibility.

Drainage is crucial to ensure good performance and long life. This is often a badly abused element of the process. Rainwater must drain efficiently: water and snow falling on a roof need to be removed efficiently (or controlled if sedum moss is employed). Water falling on a roof may be at near 0°C or up to 40°C, which can have a massive thermal shock effect and cause both stress on the materials and dimensional changes due to expansion.

Ensuring that the roof can absorb these changes in temperature over hundreds of cycles without any failures is a very carefully controlled process. Not understanding this, or changing any of the elements, will result in a shortened life.

Often there is pressure to remove the falls on a roof, or to use less realistic criteria for the rainfall figure. Both of these will reduce the cost of the roof and have no immediately apparent effects. Insulation may be downgraded so that, by calculation, it just achieves the required thermal resistance at project handover, but the performance will steadily reduce from then onwards, becoming significantly poorer in time.

The vapour-control layer (VLC) is the membrane that prevents moisture-laden air from the building flowing through the roof section, cooling as it flows towards the outside until it reaches a surface cold enough to promote condensation. This moisture then degrades the materials around it, reducing insulation performance and generating rot and corrosion over time. This is seen as unnecessary by many, and as the negative effects are long-term and almost always concealed within the construction, it is sometimes omitted. This (often hard-fought) argument for the omission of the VLC is founded in a basic lack of understanding of thermal dynamics. Most building professionals use U values as the guiding light for thermal performance. However, a U value is just a guide – a ratio based on steady-state conditions, whereas the occasions when a building is actually under steady-state conditions are very rare. Buildings are always under a dynamic flux of energy, with both inside and out rising and falling independently. Generally, the inside will follow the outside, but out of phase and at a lesser amplitude.

The energy flow across a roof is rarely as the U value would suggest. This is important as the common test supposedly establishing the need for the VLC is based on the calculated U values. It is only by experience and experimentation that we know a particular U value will generate a certain performance. This implies the need (or not) for the control of vapour across the construction section. However, the testing of this performance in real values has been left to research establishments. The skill of the specifier is to establish this and ensure the specification is correct, given the application. However, lacking hard-edged confirmation, this is an easy debate to lose when time and money are pressing.

With the need to conserve energy and carbon, there is now a chance that the performance will start to be looked at in real terms, and for the first time we may see thermal performance properly understood for everyday projects. Energy performance certificates are also a help in winning the argument that a specification should be carefully analysed before changes are considered.

Drainage is also critical, and the provision of adequate drainage, or response to the weather, is often omitted during tendering.

During the tendering process, it is often the lowest denominator that wins the day. Cross-checking the terms for performance of drainage and related items is often overlooked or is not considered important. The 'catch 22' is that the tendered figure is often within the cost plan, but the performance is of the lowest, to achieve the minimal performance. To put the required level of performance back into the tender will obviously cost more money and will therefore be resisted. It is at this point that the design professional needs to bring to bear all their skill and technical agility to explain clearly the implications of accepting the tabled tender. If, after examination of the risks, the client is still certain that they are acceptable, then clearly that is all the consultant can do.

Ensuring the specification is correct while not being overblown is critical to the next stage in the process. The impact of any particular item will depend on its size. The scale of any element and its proportion will be fundamental in the consideration given to it and the impact it is likely to have on the project as a whole. Consequently, large components need to have more time devoted to their make-up. The potential for errors is considerable and will have a proportionally significant effect on the building.

Equally, though, consideration of small but significant components can also be an issue. Specification of the correct fixings and sealants is critical. That fixings need to perform seems an obvious statement; however, the result of any failure can be a major failure for the building. Any fixing concerned with the fabric should be checked for compatibility. For example, bimetallic corrosion is a common problem that shortens life and creates secondary problems. Fixings, joints and seals are normally hidden in the completed building, and any corrosion or failure will not become apparent until it is too late.

Medium-sized components, such as doors and windows, normally suffer through use. For windows, doors and cladding panels, establishing the use pattern is critical. Often a specification centres around the number of operational cycles a door or window can achieve without significant breakdown. Lessening the specification will reduce the performance. Manufacturers' estimates of performance need to be considered in detail by the designer, and if possible verified independently.

This example can be extended to apply to nearly every part of a completed building. The practical and procedural pressures that are brought to bear may effectively stall progress from a well measured starting point.

Procurement

Ensuring the materials and assembly get to site is also a potential problem. This is often left largely to the subcontractor with little or no supervision. Potentially, a package manager will consider the details, but the room for error is ever-present due to lack of knowledge.

A roof assembly is very prone to errors, as once the finish is installed, checking the correct build is almost impossible. Drainage systems and levels are also difficult to verify until it is too late.

At completion, errors also often occur with the handover information, and we are notoriously poor at this process. However, due to the pressure of the Construction, Design and Management (CDM) Regulations, most moderately well organised projects have clear and comprehensive handover information. This is needed to confirm the product materials and systems used on the completed building, and to advise on how cleaning and maintenance should be undertaken. It is clear that, without this part of the process, life values are not going to be achieved with any certainty.

The client body and the building maintenance personnel then need to respond to these requirements. However, countless examples, even from companies with the best practice, show that little respect is given to these documents. Many projects revisited years after completion show virtually no proper cleaning or maintenance except for highly visible areas.

Cases of the wrong cleaning systems being used surface all too frequently. This is particularly obvious with finishes, although usually several years after completion a significant problem with roof or walling systems will trigger an inspection. After handover, problems of poor care may lead to roof systems having a build-up of debris, corrosion and surface grime. These result in eventual and premature failure of the drainage, causing flooding and further damage to other areas. Sometimes this may go undetected for many years until there is a significant failure. At this point, any chance of achieving the original life estimates is highly remote.

Finishes, particularly paints, will suffer considerable breakdown due to a lack of cleaning, with the top surface being eroded and microscopic pitting removing the protective layers of paint. Dirt and grime also prevent materials and seals from working correctly, as materials move in thermal cycles, acting like sandpaper and wearing down neighbouring surfaces.

By this account, our roof has little chance of reaching its design life. This is not an imaginary tale – roofs all over the country suffer the same fate – as, by implication, do all the other parts of buildings.

Judging whole life in these circumstances is very difficult. One certainty is that it will be less than the potential. Sometimes more by luck than by judgement, there are a number of factors that help:

- weather conditions are less aggressive
- some of the resilience of the original products remains
- lack of maintenance may cause largely superficial damage.

This is important, as the breakdown of one particular element will affect the others. For example, looking to thermal performance ensures the thermal stresses are kept in check, ensuring the optimum life for the roof.

At 'life end', can the roof be reused, recycled, or have its life enhanced? The chances are that it can't because a one-size-fits-all solution makes this impossible.

Tendering

The results of the tendering process will vary depending on the type of contract that is in place. With a traditional contract, it is likely to be very accurate, but for other forms of contract the emphasis will be more on price.

Additionally, issues such as the exact performance required, or the limits of the package, may leave something to the imagination. Quite often, performance levels and significant details are left for the tenderer to suggest. This often leads to deficiencies being discovered very late in the process that are very difficult to remedy. This, of course, leaves any life values way behind in the pecking order.

In order that this does not occur, or is limited, the details of the package should be established early in the process. It is important to ensure that the performance criteria are included as required, and to check the returned documents even if that is not directly your responsibility. Any deficiencies should be identified and confirmed to the client.

Any shortfall in a roof installation will be costly at some point. It is clear from years of experience that, all too often, the package for the roof is organised around limiting responsibility, or the so-called 'one point of responsibility', but fundamentally ignoring practicalities such as expertise and interfaces.

There are normally only a limited number of companies that can deliver the right quality. Identifying the right one for the job is a very skilled process. Again, this can be short-changed unless the team possesses experience in the sector.

Responses from the tendering process need accurate analysis – this affects more than just the bottom line. Have any changes, exclusions or qualifications been included? These can make significant alterations to the whole life values.

The appointment may be with the contractor or with the client. In either case, it is crucial that the details are correct – although this may seem to be stating the obvious, it is often the case that they are not. Before being appointed or ordered, the details must be confirmed and agreed by all the team. So many issues affect the roof that it is essential to have the whole team in agreement.

Ordering and availability

Ordering and availability to suit the programme are equally important. Ensuring that the roof can be delivered when required, and that the waterproof date is adhered to, is essential. However, again shortcuts in the programming often result in poor joints, inaccurate tolerances, entrained moisture, or even the wrong materials being used.

Site establishment

Before starting on site, it is necessary to have all the tests, paperwork and ordering confirmed – this is often not checked, or checked too late. Both prior to and at the start on site, reaffirmation of the specification, scope and programme dates is a good idea.

Identifying exactly what will be built may seem obvious, but unless accurately confirmed, there is always room for doubt. There are many points along the route to site at which changes, either deliberate or accidental, can creep in, mostly unnoticed. These will be too late to change unless they are picked up before the work on site gets under way.

Checking the paperwork is one aspect of this, and ensuring the items on site are correct is another. Again, the form of contract will have a substantial part to play. In any direct form of contract, management or traditional, the design team and client's representatives will be able to check these elements.

Work on site – quality control

Whatever the contract form, a quality control sample is a good means of providing a reference point. It also can be used to establish that everyone has the same understanding, and any questions or doubts can be resolved. Having several samples en route to the approved one, and labelling it the control sample, should not be ruled out. It is important that all examples of construction are covered in this way. These references are then kept securely on site. During construction, they should be used to ensure the works are being undertaken in the correct manner.

On very large projects, there may be enough resources to take the samples and submit them to accelerated ageing, to ensure the life values are in accordance with the specification. However, it is rarely possible for this to be undertaken prior to construction with enough time to correct any errors. It can, however, help with perfecting maintenance and cleaning processes, which can be adjusted depending on the results and used to enhance the life of the roof.

Completion

This is the last opportunity to correct and check that the main contract conditions are in place. On the run-up to completion, roofing needs to be tested. This should be a matter of course, with a detailed look at the joints, the rainwater collection system, flashings and interfaces. Any one of these can cause problems in the life of the roof.

Ensuring the construction team leave the site without damaging the roof as they go is essential. Last-minute work to complete the building's systems and finishing touches often result in access across completed areas that causes damage.

At handover, commercial buildings often then suffer a fitting-out programme. This will commonly require work at roof level, such as fitting additional plant, cabling or aerials. Without care, the roof will be exposed to bad treatment and damage will result. Unless well monitored this is likely to go unnoticed until a significant failure is triggered. Fitting-out programmes always cause problems, and care is needed by all concerned to minimise these. The effects on the life value of the building can be hidden and considerable.

Unless the main contract team was especially particular about recording the quality of the roof at handover, there may also be confusion over how the failure occurred. This will result in time-consuming debate and review before the defects can be put right. It will also significantly erode the potential life of the elements affected.

Ensuring that all works are carried out to avoid damage, disruption or knock-on effects is fundamental. Recording of all works should be undertaken at each stage, to ensure that any subsequent maintenance or changes start from a known position.

The prospect of thermal performance monitors will lead to another potential set of checks that require quality and performance to be clearly established and maintained.

Maintenance and use

When the building is completed, including any fitting-out, and occupied, it usually enters the longest phase of its life. If all has gone well during the construction and fitting-out, it should deliver many years of problem-free service.

However, no building can survive maintenance-free, especially the roof. The documentation should contain very specific measures, including regular inspections, certainly in the first year, to ensure all is well. These can be spread out once the annual cycle has been established and there are no local issues to take into account.

Maintenance for roofing

Inspections at handover aim to identify:

- workmanship needing remedial action
- work that is not complete
- changes from the design specification and/or quality control samples.

Inspections to check maintenance and cleaning required:

- during the first year, inspections should be at a maximum of three months to determine the rate of debris collection on the roof
- subsequently, inspections should be set to suit the roof, typically annually
- each inspection should be recorded with notes and photographs
- these should be copied to the client and the team
- each inspection should trigger cleaning and any other work identified.

All these activities should be included in the project plan agreed with the manufacturer of the roofing system, who may have other demands to ensure that any warrantee conditions are met.

All work to roofs should come under a strict health-and-safety focus. All activities must be carefully analysed and organised prior to any work being undertaken.

With all these measures in place, it is probable that the roof will deliver optimum life value. However, because roofs are arguably the most hard-working of all building elements, and they need to perform throughout their life, an interesting issue is raised that can be termed 'intermediate life'. This is because, after many years' hard service, it may be prudent to engage in some active maintenance. Rather than let the roof fail, action can be taken to refinish or recover certain types of roofing. This may involve simply adding another finishing sheet, reconstructing areas prone to failure, or changing the roof's construction while retaining all fixtures, fittings and anything reusable. For example, this could mean replacement of the top sheet of the system every 15 years. In this case, the life value of the finish is deliberately cut short in order to extend the life of the whole assembly.

Recycling and reuse

At some point, the roof's elements will all be at their life end. It is difficult to be specific technically about the point at which this occurs. This has more to do with viability, and possibly an assessment of the building as a whole rather than one element. However, it may also be that refurbishment of the whole building will trigger the installation of a completely new roof, perhaps because the waterproofing is no longer reliable, or the insulation may not be thick enough or not performing as well as it used to, or the rainwater system may not be operating efficiently.

Most elements within commercial roofing systems are difficult to reuse unless they involve metal finishes. However, most can and should be recycled.

It is at this point where the true life values may be identified. Can the materials be readily used in other products? Can they be broken down easily, or must they be burnt or dumped? If the former, can they be used safely to create energy or for other purposes?

It is clear that any or all of these actions will be decades ahead. These decisions will need to be made in a very different world from the one in which the building was designed. Almost all the criteria that were used in the original design to determine performance, and therefore life value, will have changed, and many will have disappeared altogether. It is therefore conceivable that some attributes that were thought to be perfectly reasonable at that time will be completely unacceptable, or even dangerous, at life end.

One example is that of asbestos cement sheets and roofing. Once, not too long ago, considered to be cheap, practical and relatively hard-wearing, it is now considered a hazard and must be removed under controlled conditions.

This clearly explains the nub of the whole life cost-and-value dilemma. It is very difficult to establish what is likely to happen even a few years into the future, and the effect that may, or may not, have on the life and life value of any particular element.

Our decision-making processes perhaps 30 years ago would have been seen as sensible at the time, but in the context of today they may be questionable in many respects. Extrapolating forward 30 years, we will be no better at making the correct choices unless we employ smarter thinking.

Sometimes issues may arise that make the whole evaluation seem completely pointless and reduce any efforts to prolong life, and design to ensure maximum use of resources, to the point of irrelevance. However, that does not mean such efforts should be abandoned – for every example like asbestos cement, there are countless others that will benefit from ensuring we build for as long a life as possible.

Chapter 7

History – how did we get here?

The heritage of today's procurement and design processes is at the heart of the whole life value question, and in some ways points to why and where we went wrong. The drive for ever more cost-effective solutions pushes performance to the limits, and sometimes beyond. It is time to push back a little in some circumstances to achieve something better than 'just compliant', 'just ok' or 'just legal'.

Procurement was once a simple affair. Thought of as merely sourcing of materials, it was simple, local, and above all transparent and manageable. As the ability of industry and the transport system to handle materials at an economic cost developed, so did the ability to provide more complex materials to sites at any location.

Sourcing of materials

The ability to choose any material in spite of relevance, quality or suitability may be likened to the increased numbers suffering from obesity among the general public. Nowadays people are able, and often choose, to eat beyond their needs, and as a consequence we have excessive consumption that is not good for general health. The same is true for supply chains in the modern construction industry. Irrespective of how poor a material choice is, it can be delivered and used. The ability exceeds the logic of the situation. We therefore find that, where once materials had a local and self-limiting range for their application, they can now be used appropriately or otherwise due to a lack of knowledge, or deliberately misappropriated in the light of cost, politics or dogma.

This self-indulgent principle allows materials and assemblies to be applied irrespective of the logic and fulfilment of the brief, because they may satisfy certain principles early in the building's life, in simplistic compliance. But this type of short-termism pays no real consideration to the life of the building and the cost to the building's owners, or to society as a whole. As more consideration is being given to sustainable issues, more realistic principles are needed.

It is worth considering how we got into this situation, before considering the future and how processes can be developed for greater benefit. The concept of 'built for life' is a not a new one – what is new is the conscious thought given to delivering this idea. We need, perhaps, to capture the principles and ideas behind the construction of yesteryear and put them to good effect.

A historical overview

The Victorian era was a period of rapid change and development. Transport improved and the rail network gradually achieved full coverage, enabling heavy materials to be delivered to any location. Materials could be sourced from the best quality available, and there are many examples through the Victorian age of very high-quality construction that has lasted in very good order to the present day. Arguably, these buildings are among the best examples of the principles of long life, loose fit, and low energy first identified by Alex Gordon in 1972. His paper, published before the first oil crisis, offers an important maxim that can steer us in the right direction even today. Recast in modern terms as sustainability, flexibility, and energy efficiency, it fits exactly with the principles of whole life value.

Many Victorian buildings encapsulating these values are still in use today, or have been modified to take on a completely new use, making the energy and resources applied originally exceptionally good value.

After the heritage left to us by the Victorians, the next greatest influence on our built environment in modern history is that brought about by two world wars. These two periods of horrendous upheaval, tragedy and conflict also brought rapid and dramatic change. Regarding building procurement, war had a significant effect on the ability to source materials and the methodology of construction. While old sourcing methods could not continue, new values replaced old ones. New technologies, developed largely out of necessity and applied to the war efforts, had a springboard effect in the immediate post-war periods.

In this country, scarcity of materials and labour gave rise to the prefab generation. Built for very short life to help alleviate the extreme housing shortage at the time, many such buildings have been in use nearly up to the present day, and are an example of design life far exceeding expectations. These are a great example of designs used in a different manner from that originally intended, resulting in a life increase from three to 40 years – exceptional whole life value in anyone's language. How could that have been predicted?

This is a significant point in favour of not using too prescriptive a system to value the materials and resources used, but rather concentrating on the uses, maintenance and upkeep of the building. Most prefabs survived due to the care taken of them by the occupants.

Post-war shortages created a need to look again at the way not just housing, but also many other types of building, were created. While inspiration sought new and enlightening designs, there was a lack of materials, resources, money, skill and ability. This culminated, in the 1950s and 1960s, in some very poor buildings in both design and construction. In subsequent years many have been pulled down, whereas their Victorian neighbours have prospered and are still in use.

In addition, disastrous factory-building programmes squandered resources and gave people appalling environments, creating some of the social woes we have seen last for decades, especially in inner-city areas. With the benefit of hindsight, it is clear that a balance of issues must be applied and considered by anyone developing a construction project, be it for housing, social or commercial needs.

Unfortunately, it is not clear if these lessons have been learnt. If we are to follow the much needed sustainable, low-carbon road without creating very poor environments, the lessons need to be well understood and followed wholeheartedly.

As construction blossomed, new forms of contract emerged. Clients wanted more certainty over time and money, which design teams had failed to deliver. This saw the emergence of the design-and-build contract, management forms of contract, and recently the partnering and PFI approaches.

Poorly administered and organised projects gave rise to over-runs in time and money. Clients also took the view that as they were paying, they should be in control of the money. Quite often, contingency and cost errors eroded trust. Tired of cover-ups and design errors, clients looked to other procurement systems, and quite rightly considered that they should not have to pay for such mistakes.

New forms of contract, however, took away many of the 'traditional' principles that can now be seen as the glue that held together many of the key issues we are considering. The closeness of the team resulted in more cohesive communications and better understanding. That said, more complex projects call for better, proportionately organised management, and we should be looking at the best of the old along with the best of the new.

The development of the design-and-build contract form was seen as a great move forward in this respect, ensuring – as closely as possible in construction – certainty over time and money. The formula is that employers' documentation establishes the requirements to be tendered, and the contractors bid in response. On the plus side, projects then became driven by the contractor, in many cases forward-funding schemes. Many projects may not have progressed without the contractor's input. The contractor then develops and procures the scheme. But it is important to recognise that this contract method can change elements of the design, while ensuring the overall principles are maintained. This flexibility, allowing the contractor to guarantee the time and money, is in the client's best interests, but may erode the potential performance of the building.

Reassembling the detail, or value engineering, has developed as an almost inevitable process within any project. This may be seen on the client's side as removing the irrelevant detail from the design, ensuring that the scheme includes only what he or she requires. On the other hand, it is also a process that can remove many of the attributes delivering the life and wellbeing of the building, often 'throwing the baby out with the bathwater'.

How does this arise? Primarily through the design not matching the cost plan, the client being suspicious they are being asked to pay for more than they asked for, and a general cynicism that pervades much of the construction industry.

There is a clear need for professional advisors to have the trust of their clients, particularly with whole life cost issues. Therefore the design, the analysis undertaken, and the manner in which they are presented must all be clear and convincing to both client and advisors.

Decisions made in whole life costing in particular need to be acted upon over many years so that the prescribed end result is achieved. Many clients will not be interested, or this will not be part of their business plan. However, the core of the investment strategy will rely on the issues over life costing. Many fiscal models operate on relatively short-term timeframes, which can have a substantial effect on what a client may be prepared to invest in a particular project. This may be the reason behind a team's inability to see the wider picture. The structure of the UK property market is also a limiting factor here. The predominance of developer-led property means the focus is on relatively short-term leases, with repair being structured around the lease interval rather than what is needed by the building. This results in a building maintenance, replacement and refurbishment process based not on the rational use of materials and resources, but on financial agreements that inevitably focus on short-term incentives.

While it is difficult to see this trend changing substantially, there are signs that the process may be becoming more flexible. Pressure for change is building in several areas.

The leading position continues to be taken by owner–occupier developments. For some time, the blue-chip sector has been increasingly interested in the life and life cost of their buildings, and not just in the monetary sense. Many companies are looking at the resources used and the environmental costs, and expect to consider sustainability as a main business driver. To these companies, the importance of sustainability is a boardroom position and a growing thread of influence throughout the company structure.

Many have now been operating these procedures for several years and would argue that the effects have a definite positive outcome for the bottom line. They also look to the drive for better sustainability and consideration of whole life issues to improve the company's public relations with customers, staff and business partners alike. This lead sector inevitably has an effect on the remainder of the property sector. Benefits demonstrated by the leaders are now having an impact on others more reluctant to leave the comfort of the traditional, well trodden route of a full repairing lease.

Additionally, the astute will have noticed growing pressure on the government to show it is fully committed to the green agenda, as increasing legislation is closing in on the easy options. We are seeing low-carbon and green initiatives abound – with the hope that they will result in real change for the better.

So – like it or not – the pressure is on for clients to consider more carefully the resources, materials and energy they use, and how, once a building is completed and occupied, they can be used to maximum effect and for some considerable time. Much has yet to be addressed in respect of people's use of buildings, and in balancing the need for buildings of the future to be understood and controlled as designed, rather than ignored or essentially abused.

Turning back to the past, what can we learn from previous generations? The Victorian terrace was the cornerstone of housing for the industrial revolution and for many decades. It took the form of the 'two-up-two-down' with kitchen at the rear. Built to a simple plan, it was the housing common dominator. It arose from a need to improve on the more squalid back-to-back homes that it replaced. The terrace principle was rolled out across the country, and nearly every town in the UK had its version of these buildings, providing the majority of urban housing. The principles were very simple. They did have some degree of foundations – a step up from the Georgian era. The walls were initially solid brickwork, and later moved to incorporating a cavity to improve damp resistance. The plans were simple, if cramped. The windows were sliding sash, a simple device for encouraging ventilation from one opening. The roofing was a simple pitched arrangement stretching the length of every terrace, with slates sourced from Wales.

The net result was buildings that were simple, but mostly built of very durable and good quality materials. In many cases they lasted nearly a century without too much modification. However, even the obvious addition of bathrooms and more opulent kitchens has been incorporated without too much of a problem. Many have had the benefit of reroofing, largely due to the fixings, then the structure, failing, but the tiles could go on for some time yet, such is their quality.

The lesson that can be learnt here is that a simple plan, using good-quality materials and built well, can provide stunning long life. Add to that an ability to cope with modification as standards and fashions change, and these original buildings can be regarded as exemplary value, outlasting many younger buildings. The original builders of these properties would never have expected them to survive for as long as they have, and probably this did not even concern them.

The eventual downfall of the majority of terraces has been their cramped layout, steep stairs and overall arrangement, which ultimately may not be flexible enough for modern requirements, rather than any significant failure of the buildings themselves.

This is in contrast to the Georgian town house, which in many ways is a paradox. While the epitome of style and grandeur, these houses were often built of very poor materials, being all image and very little substance. This is a great disappointment, as many would prefer the aesthetics of the Georgian to the Victorian. The need for style to be delivered by craftsmen in the Georgian house limited the ability to deliver in all respects, whereas the increasing industrialisation of the Victorian era allowed construction to be produced in volume.

Many would see the Georgian period as providing the most sophisticated design quality, but the materials they used were not to last. However – and this is why they are paradoxical – while potentially not having very long lives due to poor specification and construction, their saving grace has been style. This has attracted investment, and the deficiencies of the original have been offset by the aesthetic value. This has kept such houses valued and investable to this day.

From this example, we can see that the whole life perspective may not be due to an original design that has been given good whole life values, but sufficient character interest and kudos can radically change the whole life value. Therefore care has to be exercised when assessing whole life, which can be vastly changed by effects not usually considered as part of the original design.

The Tudor home developed from the timber-frame halls of medieval times. The craftsmanship and skills required to create the large structural frame were developed and honed in the boat-building industry. The combination of the substantial frame, the jutting to create more of the safer upper stories, and the largely thatched roofs gave a simple system of building some long-life features. Many of the examples that survive do so because of good fortune. These buildings, while having substantial frames, have two major weaknesses. They had thatched roofs, which were very prone to catching fire in an age when open fires were the only means of heating and cooking. They also rarely had any foundations to speak of and sometimes, despite the large timbers, did not behave well structurally.

While the components were solid and of quality, the build design and some elements were less resilient, resulting in poor life values. However, many were rebuilt after disasters and the frame components reused to build other houses (indeed, many of the timbers were originally recycled from ships). Many brick-clad and Victorian-looking buildings have under their skin a Tudor timber frame.

Even the Roman villa is a source of some inspiration. Over 2000 years, most of the substructure has gone, but close examination shows that the basic construction was sound and was set for a long life. These structures were abandoned after the fall of the empire, or pillaged to be used in other buildings. However, many hypocausts and delicate mosaic floors have survived, giving an indication that, even in Britain, the quality of construction was very high. Subsequent generations used the components to source their own buildings.

The essence of this historical overview has been redundancy resulting in resilience, where materials have been reused in several construction cycles.

Chapter 8

Ageing and associated factors

Central to the practical application of whole life issues is the question of ageing. If we could achieve a principle of designing or engineering out the process of ageing, whole life would take on a whole new dimension.

Above any other consideration, ageing will colour any decisions over the future life replacement or continued use of a building, or any element within it. It is therefore useful to examine what we mean by ageing, and what influence it has in any debate over whole life issues.

Ageing

Ageing is the ability of a material to stand the test of time, or the rate at which it degrades. All materials and systems interact with each other, and this interaction is often interpreted as ageing. Being in control of these processes is the goal of any designer with respect to building performance.

The importance of ageing can be summarised as:

- the factors that affect life directly
- a key to the life of any material
- the balance between use and abuse
- a true test of quality
- central to life value.

However, ageing is affected by the quality of the original specification, materials, quality of installation and level of maintenance. Examination of the properties of each of these elements and how they are affected in time is necessary to establish what exactly the life cycle is, and how we may fashion it to suit the needs of the original brief – but is rarely acted upon or achieved in a practical sense.

It is disappointing that, even in projects where life-cycle costing is present, there is very little real application. This is because analysis is based on initial cost, with very little real attention paid to the resultant cost or implications of the decisions made.

Most PFI and PPP projects are the major offenders in this regard. Huge funds have been squandered in the rush to produce new public buildings which have been cost-driven. This will undoubtedly lead to a great deal of premature failure in years to come.

The basis of our procurement industry is cost, and least cost at that, with any consideration of ageing largely ignored. This is extremely short-sighted, and results in problems even before buildings are finished. Many need further work to correct these

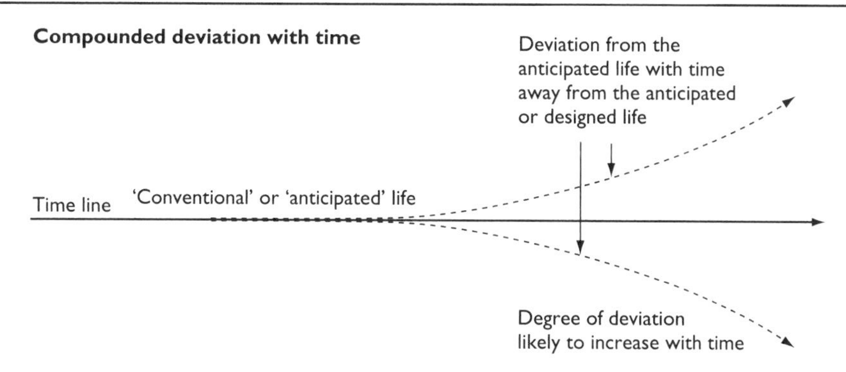

Figure 8.1 Issues expanding out as a chain of actions

errors before completion. Little is known of the maintenance costs incurred, usually because they are in a separate budget, or even under a separate contract.

Hopefully, the pressure that continues to grow in respect of sustainable solutions will result in better, more mature consideration of these issues.

This effect is made clear in the survival of many old buildings, built of high-quality materials in a robust style and with substantial levels of craftsmanship. They continue to do a useful job even when refurbished or changed beyond their original use, perhaps several times. Schools that were community centres now are apartments. How would we have applied whole life costing to these historical, classic buildings?

Where are we going wrong, and what can we learn for the future?

The fundamental issue is that initial cost, while important, is not the be-all and end-all. Consideration must be given to other factors for the design and construction team to develop a real answer that is balanced and appropriate. The real whole life cost can then be established. As resources become more limited, and proper thought has to be put into design and procurement, the need to consider these issues will rise up the agenda.

Tangible evidence for these processes is seen in the ageing of the materials and systems in the building. It is helpful to identify the component characteristics of each element.

Material characteristics and choice

Original quality, production sources, standards, type, benefit, loss of performance over time.

Workmanship

Standards, methods, care and attention, quality control, appropriate handling.

Procurement

Sources, transport, assembly, standards, quality control, understanding of needs.

Installation

Assembly, method, quality standards.

Maintenance

Planned preventative maintenance (PPM), actual methods, care and attention, rigour.

Cleaning

Actual closeness of fit to original requirements, practicalities, quality, consistency.

Refurbishment

Most buildings will need periodic refurbishment, and this should be considered from the start.

New life/demolition

At least in principle, these should be part of the overall plan, although we may not be able to see the detail when the building is conceived.

Use/maintenance

Abuse is common, and we need either to design for this or prevent it. Either way, to have any chance of success in establishing and promoting the whole life value of the project, better means are needed of promoting the principles and helping user understanding. While sensible, control of a building's use is rare. Huge resources and opportunities are wasted because of the lack of control. Even buildings that we might assume may not suffer in this way, such as schools, are nearly always badly abused and equally badly maintained, shortening their life.

Ageing as a core factor must be considered for each of the following:

- common usage
- good points/bad points
- maintenance
- abuse
- repair
- cost to client
- cost to society.

Building types

Life values will be of varied importance to various types of building. The range of building types we have is largely due to use and economic importance. However, other cultural or social aspects may also affect the life potential of a building. All religious buildings, for example, are built to an extremely high whole life value. Culturally, these are physical embodiments of religious strength, in effect saying 'my beliefs are proportionate to the quality and life of this building'.

This is the extreme case, but some correlation in others is clear. This is another aspect of whole life costing and value, driven by a number of complex issues that are not immediately qualifiable and are extremely difficult to attribute with values. Only general assumptions are going to be possible for buildings controlled by such issues.

We can, however, consider each in terms of its value to society and the value that society places on it. This social standing or ranking is the worth placed on a particular building – the value the owner or user of the building places on it, and/or the value society places on it. A garage in Chelsea may be sold for many hundreds of pounds, the same price as a whole estate in Scotland. Which will have the most long-term value or the longest life? Some will say this is entirely due to location, and in one sense it is, but equally there are social drivers that affect the outcomes. Will churches be as well cared for in the future as they have been in the past? There is some evidence to suggest they have already started to decline in importance within our civilisation.

Economic standing is possibly the core issue. Location, proximity to resources or commercial centres, is a main driver. Demand for accommodation at a popular location forces up value, but only in relative terms, as we see in any market downturn. The most popular survive; the least do not.

But historical, religious and cultural standing all give a building value beyond this economic definition. This will distort any whole life analysis. Can we create cultural value to preserve buildings? Lack of ownership, lack of responsibility, lack of feeling the need to care will often lead to devaluation of many buildings and their premature demise.

Chapter 9

The process of procurement

On many UK construction projects, the procurement team has considerable power, often to the detriment of the finalised project. The mantra will be to deliver the project at a cost – normally the cheapest possible. Cost plans that are produced quite often do not reflect the real cost of the project, and right from the outset the race is on to deliver to the cost plan.

This is never more obvious than during the procurement process. The UK industry has become used to chasing cheaper and cheaper targets, regardless of the long-term outcome. Procurement teams are, after all, doing their job. They need to find suppliers who will supply the required item at the price they determine. However, more often than not any or all requirements suffer due to price.

Most procurement teams have no idea of the technical issues involved, let alone the aesthetics of design significance. It is therefore a process akin to a lottery of performance issues. If the team is strong enough, and well organised, these points can be challenged – although the challenge is often defeated or modified – and it is the building that suffers.

So why is life-cycle costing so neutralised by money? It is largely because the issues have been taken over by that part of the industry. We are therefore left with an analysis that is largely removed from the real, practical and everyday world.

Is value engineering value for money?

Value engineering comes into play at some point in every project. The 'high altar' of cost control, 'value engineering' is a term that should be consigned to the deepest pit, along with 'mission statement' and similar cul-de-sac thinking. The prominence of this kind of collective madness should be a lesson to us all in future. Beware the management specialist offering a better way to be logical.

There is nothing resembling either value or engineering in this process. This is because most clients have a fundamental responsibility to deliver on budget to their board, financiers or colleagues, whereas designers have been seen to be frivolous and not good at delivering value. The focus on the cost is crucial, but all too often to the exclusion of all else.

Many clients even consider 'contingency' a dirty word, and that is misguided. Contingency is needed due to the industry's inability to do almost anything accurately, including the cost plan. Additionally, every building is a prototype, and while the team may be experienced and professional, many relationships and interactions will be occurring for the first time. Contingency is a planned method for ensuring that the gaps in knowledge, information and inevitable fallibility are plugged and a whole project is produced at the end. Again, there are many examples of contingencies being used inappropriately, but this is equally true of most elements in the building process.

The forces of value engineering during the procurement process will result in changes and, most crucially, any resilience within the design will be squeezed out by this process. Any initial considerations of the life of materials will be long gone by this time. This is a central issue, as the life of any material will be inherently tied up with its physical characteristics. The more these are pared away, the slimmer the chance that the anticipated life will be achieved, and the greater the chance of increased maintenance as a direct consequence.

Value engineering therefore not only is a source of increased costs during a building's life, but also substantially reduces the potential life of the project. This is seen in the substitution of materials, reducing one specification for a lesser one, removing resilience by reducing material thickness, etc. Commonly, stainless steel fixings have been replaced by galvanised mild steel, insulation reduced to the cheapest and least robust, fixings moved to the maximum centres – the list goes on.

This should be the heartland of whole life costing. With every change at this level, the life of the project is reducing, the maintenance need increasing, and the overall real value of the building becoming poorer.

Many of the issues and the problems at the core of life-value analysis occur during the procurement process. In many projects, it is here that the fundamentals of the project are decided. This is because procurement is a crossroads in the transition of the project from design to reality. Procurement is the process of turning the detail into a definitive set of orders from manufacturers and subcontractors. These will reflect the design details, specification and contract conditions. It is important to realise, though, that this is mostly controlled by price. It is the cheapest supplier who will normally get the order.

It is essential therefore that the core attributes required are built into the documents at early stages. These must be confirmed at tender and reconfirmed as the work is procured. This is where the building process starts – where the theoretical issues are turned into the practical elements that will be assembled as the building.

Often the process starts to distort the objectives of the design. This may be for a number of reasons, some of which are perfectly understandable:

- the design is not fully thought through
- elements of the design cannot be achieved within the cost plan or any realistic cost
- construction sequencing and practicality changes the design
- 'better' ways of achieving the result are offered
- clashes with regulations or standards
- clashes with other elements of the project.

All of these may result in changes during procurement. The danger is that their effects may not be obvious or apparent for a very long time.

Clearly, there are similarities with the tendering situation. However, tendering is competitive, first past the post, whereas procurement (while in some definitions it includes tendering) is a broader, more general process. It can therefore have a greater effect and cause more disruption to the project in the long term.

The purpose of procurement is turning the project into a reality, ensuring the components arriving on site are fit for purpose and are delivered at the right time. However, there is serious potential here for the life-value qualities to be considerably affected These can be complex and far-reaching, but worst of all, not obvious for many years.

What elements affect life value?

Often the passage of documents from design to confirmed order may result in a reduction in:

- material quality
- material thickness
- material protection (permanent and temporary)
- fixing quality and type
- quality control of products
- quality control on site.

These do not appear as obvious issues, but as changes buried deep within the documents. Often these are substitutions of specification, or may be uninformed changes, or assumptions that one specification is the same as another.

One significant problem here is the maxim of 'equal or approved equal' or a similar description. This phrase is used to allow the substitution of a material or product that the contractor can fit into the programme more efficiently, or simply is available (see Chapter 5).

The term 'approved equal' and a myriad of variants has come to be something of a spectre within the construction industry, much maligned on all sides. It is important to recognise that this is a legitimate and very useful device. The ability of the contractor to add his or her skill in ensuring that parts of the project are appropriate in terms of performance, but also are ready to be included in the build at the right time, is a very powerful process. Furthermore, the contractor may also use some commercial power to get all this at the best possible price.

However, it may also mean that the procured item may be woefully inadequate in a number of respects. The stretching of the term 'equal' is sometimes taken to ludicrous levels. This is because it is very difficult to prove something is equal in all respects. Different products and materials work in many different ways, making the analysis of their performance very complex. The original specification and details may not have made explicit all the necessary criteria; in fact, they are rarely identified to such a degree that the definition is clear.

In a complex world, designers rely on shorthand expressions, or on the manufacturers themselves. This leaves plenty of room for interpretation from the outset. For a designer to lock down the requirement would normally take considerable time and effort. Normally neither is available, so from the start of the process the clarity is not there and the implicit performance is lost. This is crucial in reducing the life value.

The issue of 'equal or equivalent', or any other phrase attempting to describe the principle of two entities that are, for all practical purposes, the same and behave the same, is usually very difficult to determine. This will depend to some degree on the level of detail available and the level of understanding by the teams involved, and may in the end be a case of considered judgement. This is clearly difficult when looking at a series of characteristics that may be different, or based on different issues, to establish if they are equivalent. This is a skill based partly on experience and partly on a determination to review as much of the data as possible prior to coming to a conclusion. But time is often of the essence, and therefore any review cannot be totally exhaustive or conclusive.

Maintenance

Maintenance is much maligned and often ignored. While the best plans may be laid for the generation of the project, construction and completion, this may all be to no avail. There is a clear relationship between the rigour of the maintenance programme and a building's life.

This, in turn, is related to the eventual cost of the building. There is a complex relationship between the cost of acting and the cost of not acting, increasing with time.

Chapter 10

A material matter

Do's and don'ts for each material process and relationship

It is essential that any analysis of whole life value considers three basic ingredients: quality, content, and choice of materials. These are the cornerstones of all buildings, and the culmination of design, specification, procurement, construction and assembly.

Looking in some depth at the constituent elements of the common groups of materials can help identify some of the important issues:

- concrete
- stone and masonry
- steel
- other metals
- timber
- other natural materials
- man-made materials.

Concrete

Concrete is a very old material and is very common across the whole built environment. Its ability as a structural material – one that can readily be formed into almost any shape – has given it universal appeal for many centuries. It can be used on its own, or to bond other materials together, usually masonry.

Underlying its common usage is its very long life. If well made in the first place, it can survive for many decades, still performing to the same level as when it was first cast. Its long life is due partly to its ability to resist environmental effects, and partly to its built-in resilience, largely because it is nearly always very dense. It can therefore be very reliable.

Concrete is not perfect, and has to be carefully mixed and cast. It is a high-energy material largely due to the cement content. However, decades of reliable performance can be expected from this material.

Problems for the designer stem largely from concrete's use with other materials. Many designs use concrete in conjunction with finishes of stone or brick, or merely as the skeleton behind a more acceptable façade. However, as a material concrete ticks most boxes and can be used in an astonishing range of applications.

Summary of core properties:

- flexibility good
- robustness extremely good
- care needed in specification moderate
- care needed in detailing moderate
- care needed in maintenance low
- environmental qualities moderate
- reuse achievable good
- recycling achievable good.

Stone and masonry

Stone, being a hard, natural material, is as durable as concrete, but is less flexible as it is mined not manufactured. Nonetheless, its ability to perform for countless decades without substantial problems is the reason why most old buildings are made from it. Essentially, it is robust and resilient.

Masonry is a relatively new application – a blend between stone and concrete, with the benefits of both. Throughout history it has been used as the default walling medium. Many examples exist of very long-lived masonry. It is obvious that this construction material will pay back the investment, hence its popularity.

Summary of core properties:

- flexibility good
- robustness extremely good
- care needed in specification moderate
- care needed in detailing moderate
- care needed in maintenance low
- environmental qualities moderate/good
- reuse achievable good
- recycling achievable good.

Metals, on the other hand, are nearly as popular and have also stood the test of time. It is important to recognise that metals are probably the most flexible of building materials, used for structure, cladding, roofing, fixings, finishes and fittings.

Steel

Next to aluminium, the most common metal used in buildings. It is strong, stable and malleable, therefore its use in structure, major support and fixings is obvious. It is also used occasionally as cladding. However, it bonds readily with oxygen, making it vulnerable to corrosion. Steel should therefore be used with caution. Measures are needed to ensure it is protected, and the life of the material is largely related directly to how well this protection is applied.

Summary of core properties:

- Flexibility good
- Robustness extremely good

- care needed in specification moderate
- care needed in detailing high
- care needed in maintenance medium
- environmental qualities poor
- reuse achievable good
- recycling achievable good.

Other metals

Aluminium

Aluminium is by far the most common metal used in construction. It has limited use structurally, but due to its malleable and light nature it is used across a wide range of applications, including framing, cladding, fixing and fenestration. Its relative strength in comparison with its weight makes it ideal for these types of secondary components. Unlike steel, it will naturally oxidise and stabilise, and whereas rust will eventually degrade steel to a substantial level, this is not true for aluminium.

Aluminium can also be relatively easily protected, either with a variety of paints or anodising, a system to bond a chemical finish into the outer molecules of the metal. This is a very robust finish indeed, and can provide protection for decades without any reduction in performance, although it can change colour.

However, aluminium is relative easily damaged in use, therefore its potential life can be directly related to how well it is cared for and maintained. With good care, aluminium is very long-lasting, and can be in service for many decades without problems.

Summary of core properties:

- flexibility good
- robustness moderate
- care needed in specification moderate
- care needed in detailing high
- care needed in maintenance medium/high
- environmental qualities good
- reuse achievable good
- recycling achievable very good.

Cast iron

Cast iron is little used these days, and has been largely supplanted by steel. However, it is interesting for its place in history as the first metal to be used universally in structural applications. Consider the wealth of structures made in cast iron that are still standing and still in use. But in the modern world it has been overtaken by other materials that can easily outperform it.

Copper

Copper is one of a series of metals used in construction that is relatively soft but durable as a finish. The ability to use its ductile nature to form it into sheets has made it ideal for roofing. Like aluminium, it will oxidise readily and will stabilise if left undisturbed. The attractive green colour of the oxidised metal is used to good effect. Copper can be relied upon to last for a very long time, and may be recycled.

Summary of core properties:

- flexibility good but limited
- robustness extremely good
- care needed in specification moderate/high
- care needed in detailing high
- care needed in maintenance low
- environmental qualities moderate
- reuse achievable very good
- recycling achievable very good.

Lead

Lead is very similar to copper – it oxidises to a dull grey rather than the exciting green of copper, but has the same durable qualities.

Summary of core properties:

- flexibility good but limited
- robustness extremely good
- care needed in specification moderate
- care needed in detailing high
- care needed in maintenance low
- environmental qualities poor
- reuse achievable very good
- recycling achievable very good.

Zinc

Zinc is also a material with long-life, ductile qualities that enable it to be used as roofing or cladding, with confidence that it will last for decades.

Summary of core properties:

- flexibility good but limited
- robustness extremely good
- care needed in specification moderate
- care needed in detailing high
- care needed in maintenance low
- environmental qualities poor
- reuse achievable good
- recycling achievable good.

Titanium

Titanium has only recently seen use us a cladding and roofing material. This is because it is relatively rare and difficult to form. Its hardness, however, means that it does not oxidise readily and will maintain its shiny appearance, making it seem ageless. This characteristic makes it desirable as a cladding material, giving buildings guaranteed age-cheating ability.

Summary of core properties:

- flexibility good but limited
- robustness extremely good
- care needed in specification high
- care needed in detailing high
- care needed in maintenance low
- environmental qualities poor
- reuse achievable good
- recycling achievable good.

Timber

Timber is a natural resource and therefore should be sourced, used and applied wisely. The designer should understand the production mechanisms and characteristics of the material, and apply these when searching for the appropriate material to be used in a project. Application of the correct characteristics is one of the keys to specifying correctly.

Timber is the other major building material after cementation materials and metals. Undoubtedly the oldest building material, its ability to be formed into a huge variety of shapes and its strength make it ideal for framing, cladding, roofing, and doors and windows. Timber can be used for all or part of buildings in a way that is not matched by any other material. It also has insulating properties and derives from a renewable natural resource.

It does need to be treated with care for long life. It is vulnerable to attack by pests, rots caused by excessive moisture, and fire. This is a material that needs to be specified carefully, installed correctly, and cared for if long life and a return on the investment is to be ensured.

There is a vast range of timbers that can be used in construction, making the best fit to the task in hand a challenge. Broadly, they can be subdivided into timber for:

- finishes and fittings
- carcassing and framing
- structure
- cladding.

Hardwoods and softwoods can be used for most of these applications. However, these days it is essential that the specifier uses sources that are sustainable. The best method of achieving this is to use the Forest Stewardship Council (FSC) or Programme for the Endorsement of Forest Certification (PEFC) systems, that try to ensure timber is from a reputable source.

Summary of core properties:

- flexibility extremely good
- robustness very good
- care needed in specification high
- care needed in detailing high

- care needed in maintenance medium/high
- environmental qualities extremely good
- reuse achievable very good
- recycling achievable very good.

Other natural materials

Other natural materials are used in significant areas in buildings; however, for the most part they are secondary, used as fixtures, fittings or finishes. Some important examples are identified below.

Terracotta

Terracotta is used in roofing and cladding systems as well as some types of masonry.
Summary of core properties:

- flexibility good
- robustness moderate
- care needed in specification high
- care needed in detailing high
- care needed in maintenance medium/high
- environmental qualities good
- reuse achievable good
- recycling achievable good.

Rubber

Rubber is used in flooring and seals. As a flooring material, it is exceptional and can outlast the building surrounding it.
Summary of core properties:

- flexibility good
- robustness very good
- care needed in specification high
- care needed in detailing moderate
- care needed in maintenance medium
- environmental qualities extremely good
- reuse achievable very good
- recycling achievable very good.

Wool

Wool is used for its outstanding insulation qualities. It resists moisture and maintains it insulative value for many years, outperforming man-made insulations.
Summary of core properties:

- flexibility moderate
- robustness very good

- care needed in specification high
- care needed in detailing moderate
- care needed in maintenance low
- environmental qualities extremely good
- reuse achievable very good
- recycling achievable very good.

Man-made materials

Glass

Glass a very common building material. Its main use is for windows, cladding and roofing. It is also used as a decorative finish and as shading. Glass is a unique substance in that it is nearly transparent, very strong in compression, and very weak in tension and shear. It is malleable and can be formed into a wide range of finished products.

It basic characteristics have been known for decades. However, technology has increasingly pushed the limits and designers have been keen to take advantage. Glass is being used in increasingly sophisticated applications. The composition, size of pane, and performance have increased in very rapid order.

However, the application of advancing technology has its price, in particular the ability to understand the life of these products and the possibility of early failure.

Glass may be specified in a number of ways:

- pane size
- strength
- transparency
- translucence
- optical quality
- thermal resistance
- thermal gain properties
- security
- robustness.

Glass is crystalline in normal conditions, but never quite sets, and over time it will sag. This effect is minimal, but may limit the long-term life of the material. Of more significance is the thermal stress a glass panel will undergo during construction and everyday use.

Depending on the required performance, glass can be strengthened or toughened, laminated, or a combination of these. This will result in a higher performance, but it will be at a higher cost and potentially shorter life. Strengthened glass has two potential failure points. First, the process builds up stresses in the glass, and the release of these forces for any reason will destroy the glass. Second, the strengthening process can encourage crystals of impurities, known as nickel sulphide inclusions, to form. Under certain conditions they will change shape and, in doing so, shatter the glass. For any toughened glass used in construction, the potential for spontaneous breakage must be borne in mind. It is generally accepted that this can affect up to 4% of all toughened glass.

Glass potentially can be regarded as a very long-life material. Many examples exist of glass dating back hundreds of years. However, it rarely survives unbroken, and if it does, history shows us that in comparison with today's performance it is very poor, its relevance being largely historical and aesthetic. Although its worth may be greatly diminished in performance terms, other factors may well escalate its life-cycle value disproportionately, as in the case of stained glass windows.

This raises a key additional factor when considering the life and value of any material. While it may perform to the same degree as when originally made or assembled, its performance value may diminish considerably over time. A sixteenth-century leaded light window is extremely attractive to the modern eye. When it was made, it was a high-tech piece of construction, a combination of an ability to form lead in relatively slim sections and a limit on the size of glass panes that could be created, combined with techniques of colouring and painting glass. But while many have survived intact to the present day, the security, strength, thermal and acoustic performance they offer is very wide of the mark that would be acceptable in today's buildings.

From today's perspective, with the pace of technological development, the chance of a material's life being cut short because of fashion or performance is very high, in fact probably higher than any other cause, making any estimate of life-cycle value extremely difficult to determine.

Maintenance of glass is also a crucial area. Glass relies on regular care because of its relatively fragile nature, and because in most cases it is held in place using frames, seals or fixings. Loss of performance by any of these will cause significant failure. Additionally, the primary function of most glass is to allow light into the building. Any dirt will diminish this performance, and will also have a detrimental effect on the glass. Cleaning is absolutely proportionate to the long-term life of the glass.

Glass is not affected by many of the environmental agents that attack other materials: radiation and pollution have minimal effects. However, wind, strong rain and snow can be catastrophic.

Summary of core properties:

- flexibility poor
- robustness mixed
- care needed in specification very high
- care needed in detailing high
- care needed in maintenance high
- environmental qualities moderate
- reuse achievable moderate
- recycling achievable good.

Insulation is also a significant material, provided in many forms, and very relevant to the long-term performance of any building.

Mineral rock fibre

Mineral rock fibre is a good insulator, inert and non-combustible. It is also relatively easy to recycle. Nearly all insulation materials are quite fragile and therefore susceptible to damage. However, they are normally built into any construction, and damage will occur only during construction maintenance. Life can be severely curtailed if moisture is allowed in proximity to the insulation. This is normally caused by failure of waterproofing, or due to a lack of vapour control and the onset of condensation. If these are correctly controlled then this material will last a very long time.

The limiting factor with this and most other insulation installations is that the thickness originally installed rapidly becomes less than the required standard.

Summary of core properties:

- flexibility poor
- robustness mixed
- care needed in specification high
- care needed in detailing low
- care needed in maintenance low
- environmental qualities moderate
- reuse achievable moderate
- recycling achievable good.

Foam insulations

Foam insulations such as polyisocyanurate (PIR), polyurethane (PUR) and other formulas are commonly used materials, due in the main to their low cost and relatively high performance based on insulative performance versus thickness. However, they are inflexible and hard to reuse or recycle. Foam insulations bound up in concrete constructions should have a life equivalent to that of the concrete, making them viable for long-term performance. They are limited only by the relative worth of the thickness that is used.

Summary of core properties:

- flexibility poor
- robustness mixed
- care needed in specification high
- care needed in detailing high
- care needed in maintenance low
- environmental qualities low
- reuse achievable low
- recycling achievable low.

Plastics

Plastics and man-made composites have seen steady acceptance within buildings. They are used mainly for fixtures and fittings. Some sheet material in the form of polycarbonate is used, mostly for roofing and trims, finishes and fittings.

Many common plastic components involved in construction use polyvinyl chloride (PVC) and plasticisers to counter the effects of sunlight and time. These leach out into the environment and the components become brittle, normally resulting in failure.

Summary of core properties:

- flexibility of shape and form good
- robustness mixed

- care needed in specification very high
- care needed in detailing high
- care needed in maintenance high
- environmental qualities moderate to poor
- reuse achievable very mixed
- recycling achievable very mixed.

Composites

Increasingly composite materials are being developed for use in buildings. These have very high strengths and low weight. These characteristics have seen their use in the automotive and aerospace industries become very common. Such materials are now being applied increasingly in the construction industry, especially where high-value projects allow their use.

Most applications away from construction have been relatively short-term uses. It remains to be seen if the faith placed in these materials will be rewarded with long life. History is littered with examples that did not stand the course of time, but modern materials science has a better understanding than ever.

Summary of core properties:

- flexibility poor
- robustness mixed
- care needed in specification very high
- care needed in detailing high
- care needed in maintenance high
- environmental qualities moderate
- reuse achievable moderate
- recycling achievable good.

Life required

Determining the life required is probably one of the most challenging aspects for both designer and specifier. The following issues must be clarified and agreed with the client, and explained clearly to the project team and construction professionals.

The budget

Obviously every project must have budget and a programme. Working within the budget is one of the most important disciplines for the designer. Working from an outline budget to a more defined one is the natural progression of every project.

Getting to grips with the cost allowed for a particular element can be difficult, especially if there are options. The only way of progressing is to carry out a number of iterations for a given element to see what is the best fit; this is frequently a balancing act.

The brief

It is important to ensure you are in sympathy with the brief. By the very nature of design, it is important that the designer brings innovation and excitement to the project. But to grossly exceed the brief without bringing the client with you, or to promise undeliverable concepts, will soon lead to trouble.

The client

Never forget that the client pays the bills. It is therefore very important to ensure the client is informed and clear over the development of the scheme.

The environment

It is also important to consider the environment to which the material is going to be exposed throughout its life. Traditionally, environments are broken down into:

- sheltered
- moderate
- exposed.

The location of each element needs to be considered – is the material hidden, exposed or prominent? Is it decorative, secondary or structural?

Maintenance of materials

Probably of most concern is maintenance. This is the most abused of all issues in completed buildings. It is rare that a building is maintained in the way it should be. Even with so-called planned programmed maintenance, with quality assurance systems, this part of the building specification is generally not applied as it should be. The designer can offset this to a degree by applying more robust materials, while remembering that they will have to go through the hoop of value engineering.

This is also crucial when considering the whole life cost. Most theoretical systems naively allow for optimum maintenance, which is a gross misconception. Designers need to allow for the maintenance being poor from the start. Additionally, planned significant repairs, for instance the replacement of a top sheet on a roof, rarely happen. This is not a real cost situation, but a distortion caused by common acceptance, which is why any wholly monetary whole life cost analysis is fatally flawed.

Sourcing

Matching the quality required and the specification is the key to sourcing. Ensuring that the material can deliver what is required is critical. This may often be out of the designer's hands; however, the detailed specification should spell out exactly what is required and leave little in doubt. Especially with timber, checking the environmental source (FSC, PEFC) is essential.

Specification

The specification sets out what is required: what type, what sourcing, what quality, in terms that are definitive. Workmanship standards are also defined. Reference materials and documents may be used, however excessive use of documentation has a negative effect, and these may not be used unless a dispute occurs.

Construction

Quality and care in construction are essential, as these will have a significant effect on the life of any material. An incorrect sequence of installation or incorrect placement can cause considerable long-term, hidden effects, for example incorrect application of the vapour-control layer, or incorrect use of fixings causing a dissimilar materials bimetallic effect. Correct placement results in materials remaining balanced and performing to their optimum. Poor skills on site often miss this issue.

Construction techniques of the past, due to years of trial and error, were based largely on the principle of good quality materials used in a way that was in harmony with their performance. The need to achieve better performance with materials of lesser quality has seen lifetime periods reduced and serious defects increasing. Some analysis has been done in this area, but real, practical data are scarce except when buildings fail or are demolished, or alterations are undertaken.

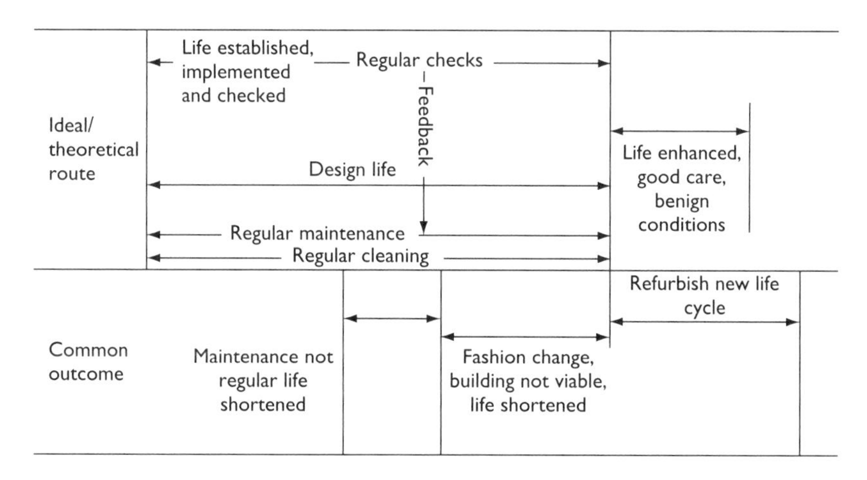

Figure 10.1 Construction and maintenance issues

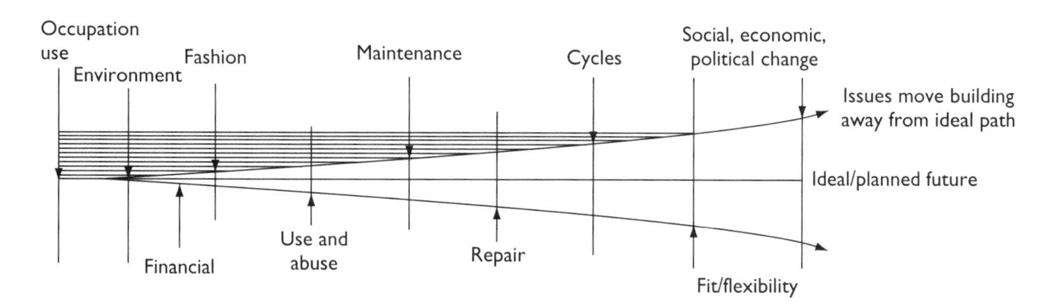

Figure 10.2 Life issues 'move' the Building away from the planned path

Construction processes include transport to site, storage on site, placement at the work face, and installation. All these processes can cause damage, generate errors, and start abuse of the material.

Additionally, there may be some time before the work is complete, with any particular element remaining exposed or in a situation that may cause long-term harm. For instance, due to the need to progress a project, work is often undertaken using water-sensitive materials before the building is water-tight. This always results in some unseen damage that will reduce the life of the elements affected and make any judgement on their likely life completely irrelevant. A clear idea of the works on site, or an assessment of the likely issues, is needed in any assessment of life cycles.

Installation and maintenance

Recommendations should be spelt out so that there is no doubt what is required. Ideally, the effects of not undertaking the prescribed work must be described clearly, giving sound reasons and the potential effects of poor attention or neglect.

- Programmed, planned maintenance is an effective way of ensuring that the life of a material is ensured. However, it needs to be undertaken correctly.
- Preventative maintenance monitors the effects of ageing and puts in place corrective measures to maintain the life of the building.

Maintenance will be of benefit only if performed correctly: investment in good quality materials, and care of them, is essential.

Chapter 11

Construction methods

Construction methods vary, and will change the performance that is ultimately achieved. Performance is the key point with regard to specification of elements and materials. How can we establish the level of performance to be expected, that will improve a building's life values?

Several measures give the specifier some confidence as to the eventual outcome; however, these can be significantly altered by issues outside the designer's control, and this will have a central role in determining the life of any element.

What is performance?

Performance is the ability of a system or material to maintain the required level of properties for a given time. Most of the time, we try to over-specify and to build in extra performance to ensure the materials are robust enough. This actually means that more resources and cost are spent to ensure the life of the materials is maintained to an acceptable level. This is important because it starts to close down the level of unknown issues, those 'unknown knowns'. We need to be clear, or as clear as we can, on the expected performance and how that is to be achieved. Understanding this as a subtext to the design procurement and construction process is crucial to the construction professional.

The significant question is, how can performance be measured? There are a number of ways of measuring or predicting the issues surrounding performance.

Defining performance

- Specification levels.
- Ring-fencing the process.
- Value engineering (possibly – as long as there is understanding of the benefits of tolerance specification levels and quality).

Controlling performance

- A measure of quality and tolerance needs to be established – often this is achieved through experience or identified intuitively.
- Comparison of many examples and previous projects may be useful.

Context of performance in the real world

This is the most complex of issues – that is it is rarely recognised or considered is testament to the degree of difficulty. The key underlying issues involve all aspects of decision-making in combination with the level of knowledge concerning the materials and detailing – in short, the design.

Design is a complex process

There are many issues that need to be considered, synthesised and resolved. Design is many things to many people. Aspects such as practicality, taste and appropriateness are the battleground for decision-making. Taste will be affected by fashion, context, cost and heritage. What is popular design, and what is outlandish design? Today's pariah of the design world can become mainstream tomorrow, and *vice versa*. Trends may be revisited, or have a comeback in popularity.

We also have distinct style periods, although these have been better defined in the past than in recent years. A design consensus results in a level of popularity that may catapult a particular language into a style period – some would also call this fashion, but it has a real effect on performance. Popular fashion may cause projects to ignore performance criteria to the point where the logic of the decision-making process has completely broken down. This can be a huge dilemma for designers, and balancing the freshness of leading-edge designs with the ability to make a building robust and to include the appropriate level of performance can be a real challenge. Disappointingly, the outcome is often style over performance.

Good design also encompasses the fundamental requirements of the client or their brief for the purpose of the building. Great design will do all this, and add esoteric values and qualities that will take the building beyond its original objectives and performance levels to create something more. Where this succeeds, it gives a sense of wellbeing and an uplifting quality that will inspire a sense of wonder in the viewer. This is the most complex of all the components of design.

It is important to recognise that behind this embodiment of detail is both technical achievement and, married to that, whole life issues. The technical appropriateness of the design will be, in part, the selection of the materials, their assembly, and the manner in which they will perform over many years of service. This is a very subtle process which is rarely considered systematically in any real sense, and one that needs better consideration in striving towards defining appropriately integrated whole life costs.

Central to this issue is the difference between value and cost. The cost of something does not necessarily reflect its true value, and *vice versa.* This must be recognised when considering the balance of quality and workmanship with performance and design.

The example of cladding

Considering performance issues from the perspective of a cladding design may be helpful. Cladding is a key issue when considering a whole life approach – it is complex from both a performance and a design perspective. The cladding of a building is probably its most critical element after structure. It is therefore important that it is considered carefully. Unfortunately, on most projects this does not happen. Cladding provides primary protection for a building (other than the roof). It controls the rainfall, sunlight, ice, snow and wind. Therefore cladding components must be reliable and perform for many years. How do we come to a realistic assessment of cladding and its potential life?

We have seen earlier how quality procurement, installation, workmanship and maintenance all play their part in general terms; now let us consider these in detail for cladding systems.

The cladding system has to incorporate several fundamental characteristics without which it cannot deliver the performance required. Each of these needs to be understood by the designer and identified to a fine level of detail. The information from the designer must enable all others in the team to complete the construction to match the original proposals.

Movement

The ability to allow for movement is crucial to any cladding system. Thermal and structural flexing, if not allowed for, would cause the cladding to fail in very short order.

- Thermal movement – all elements will expand and contract as the temperature fluxes. The design must ensure this can take place without lasting damage to any components. This must continue to work for many thousands of cycles.
- Structural movement – results from various pressures on the building loading, from wind, occupants, and possibly geological forces.

Joints

Joints are highly important as they are the obvious weak point. Joints may be designed to be open, mechanical, fixed, or sealed with caskets or applied sealant. In all cases they must allow for thermal and structural forces. They must also allow for repair and maintenance.

Open joints, such as those in cladding panels in rain screens, allow movement and other effects by careful design. The creation of a balanced air cavity behind the cladding means the joints can be left open without serious weather intrusion. Therefore the ageing effects of thermal and structural forces are largely removed and the designer can be confident of long-term performance.

Mechanical joints rely on fixing components either to each other, or by means of a cover strip. This is more complex, as movement, fixing types and position of the components can all cause failure. Each individual component needs careful engineering and installation to ensure long life. Test data on the performance will help the design team. Very often, the largely hidden nature of the subtle engineering will be lost on site, and errors causing premature failure will occur.

Sealants can be used to good effect as they are largely formless and can achieve high strength in a simple installation. Best known is the silicone glazing technique. This has been used successfully for decades, and many examples exist that are over 40 years old.

Sealants can be broken down by radiation from the sun, and stressed by thermal and mechanical effects. Sealing materials modified by heat or chemical reaction can be effective, but are also subject to radiation and thermal stress.

Drainage and weather-proofing

It is very important to ensure good drainage. Cladding should never have any areas that could trap water, and free draining is essential.

This is regularly overlooked, and this is often a prime reason for premature failure.

Cladding must always be designed to allow run-off of rain or snow and ice. It must also avoid build-up of debris. This is not considered enough in design and construction. The orientation and shape of finishes, joints and connections should be smooth and should ensure that particles are naturally flushed down the cladding by rain and wind. Systems that retain a build-up of debris should be avoided. The exception to this is green and possibly brown claddings, but these are specifically designed to control the movement of particles into the drainage system. Run-off is important to avoid wholesale cleaning of the cladding, exposing personnel to unnecessary risks. Design has to be safe as well as appropriate.

Chapter 12

Observations outside the world of construction: whole life value and cost

Linking decisions to outcomes is possibly the most difficult task in the quest to establish the success of whole life costing measures. It questionable whether any real evidence exists that this ever succeeds or has any real benefit. The result of actions will not be truly evident for many years – in many cases, decades.

Success, in whatever form it takes, may differ from one person's perspective to another's. This is certainly the case across much of the built environment. And on many occasions we put effort and resources into objectives and activities that in the end deliver little or nothing. This is wasteful of human and natural resources, and clearly cannot be sustainable in the future.

This is less of a problem in manufacturing, where a specific product is being produced. Possibly the construction industry can learn from the issues here. When developing a product, it can be prototyped, road-tested, and even destruction-tested. This will give the designers a clear idea of its performance, durability and assembly quality. Taking it to market can then be a well planned process. Of course, this is a simplification, and there are just as many pressures in manufacturing as there are in construction, and many examples of this process going very badly wrong. But in general, especially since the advent of computer-controlled manufacturing, significant errors are rare.

This focus on the product can enable whole life values to be determined very clearly. Certain manufacturers have life values as a unique selling point, and some even make it their primary message. In nearly every sector, there is at least one manufacturer making the claim that they produce products that have durability, reliability and long life (some more justifiably than others). There is a market for products that have proven reliability, performance and an almost guaranteed life. End-of-life criteria are spelt out equally clearly, mainly as a result of increasingly robust legislation to ensure products are recyclable.

These days, especially in consumer goods, whether or not these claims are borne out is easy for all to see – via the internet, the performance of almost everything is itemised, analysed and endlessly debated. For those who care, consumers have provided a wealth of research and development.

Clearly very little of this sort of information is available to the construction industry. Perhaps in future we will see the development of these sorts of online forums.

What can we learn from manufacturing?

Repetition

Construction projects are normally one-offs, created by a one-off team, and in specific circumstances that relate just to that project. Construction is normally carried out on the basis of 'right first time'. There is rarely an opportunity to go back and revisit decisions, and rarely an opportunity to test all the issues. The best we can hope for is that elements of the design can be tested and analysed separately.

Time

There is very little time or resources to test, check or analyse in the way many manufacturers do. By comparison, everything is more time-pressured and consequently less well resourced.

Prototyping

Research and development is a fraction of that used in manufacturing, where iterations of part or all of any product can be assembled and tried, and an understanding of their strengths and weaknesses established.

The construction equivalent is to ensure that as much of the project as possible is originated in the factory. These days, almost everything can be made this way. The Japanese housing industry has been producing a huge range of houses in this way for many years, but this is still an exception in most of the world. In moving to the factory, we have a standard that can be checked and assessed, and many of the life-value issues can be identified. But how this information is used is very much open to debate, and real benefits from this still seem to be elusive.

Large elements of a building can be from tried-and-tested backgrounds and can greatly enhance the potential of a project to stand the test of time and deliver the expected life. But often the potential of these elements is diminished as we cannot bring them together accurately or efficiently as a whole. The weakest elements are the items that have to be undertaken on site. Working to remove these helps significantly in ensuring the designed performance equates to that delivered.

End of life

When a product comes to the end of its life, it is becoming the norm that it should be recycled or reused. There is universal acknowledgement that products must be made use of at the end of their original life in order to maximise the original investment.

We need to consider the same for buildings; however, attempting to establish this as reality is extremely difficult. Again, a building is far more difficult to treat in this way than a product.

We also have different drivers. A product may be seen as something individual, something repeatable and often, unfortunately, throw-away, although this approach is increasingly unacceptable. A building, on the other hand, is regarded as a permanent asset (even though many are not). Manufacturing can easily replace elements or whole products that are found to be deficient, whereas buildings that have problems can be extremely difficult to put right. Some problems may take years to remedy. Unless there is a shift in this poor defect level, we will not see whole life approaches being adopted.

Chapter 13

Fashion

At first glance, fashion is an unusual item to feature in a discussion about whole life value. But it is probably one of the most important issues – and may even be the most important. Because it is not derived from technical issues, it is not based around any mainstream construction or material world logic. It is driven by social interactions and collective considerations from the wider world. From time to time, fashions do stem from technical advances (such as the iPod). We occasionally take up a change of direction and embrace it, making a significant change in the way we view and use the physical world. There is a need to consider the wider issues concerning fashion and its effects on buildings.

Fashion often results in a changing need in the built environment. This will make some buildings redundant, sometimes dramatically, but generally by stealth. Small changes, but critical ones, devalue carefully crafted designs, making any attempt at long life seem wholly irrelevant.

We need to learn from the way Victorian buildings, built to a completely different ethic from current practice, are able to be modified, albeit with a degree of compromise. It is interesting to reflect whether any of our current generation of buildings will exist in 80, 50, or even just 20 years. Some will by luck; but many will not have the same ability to adapt or to be seen as useful.

These shifts are brought about through changes in the deeper elements that affect day-to-day economics: the way we plan and shape society, and the way the politics of the everyday has an effect on the material world.

Often buildings may have their anticipated life partly or completely cut short due to a shift in fashion. This can be direct – for example, the demise of many railway stations as a result of the Beeching Report; or indirect – as in the increase in mobile phones seeing off the phone box, but will ultimately have the same effect. Perfectly viable buildings, with many years' good service, are brought to a sudden end. The key issue is that we have very little control over how and when fashion will strike, and what effect it may have.

Fashion can affect the master planning level – the growth of out-of-town shopping centres; or it can make small changes that are nonetheless significant enough to considerably affect life-value issues – television saw off the cinema until it reinvented itself.

The direct influence of fashion is seen in finishes and performance. Over very short periods, we see fashion support or alienate everyday materials, with major effects on the commercial value of a building, and therefore its life. Additionally, the ability of a particular element to perform to a particular level can also be seen as fashion. We embrace or reject at a very whimsical level. So the render or timber cladding so popular at one moment can bring on the early demise of a building: while the materials themselves, and the building's functioning, may be perfectly acceptable, its commercial value may be reduced, resulting in early demolition.

Several years ago, double glazing was seen as an important improvement over single glazing, ensuring energy conservation, avoiding condensation and improving comfort conditions. However, only a few years on, that advance is no longer good enough, and enhanced performance is required to be several times better, prematurely bringing many installations to an end.

Is there anything that can be done about this? While fashion is notoriously difficult to second-guess over the long term, there are some principles to follow that ensure some degree of stability.

Aesthetics

Ensuring that the aesthetics are within the broad level of accepted values is important if long-term life is required, prompting a conservative or classical approach, on the basis that extravagant or outlandish designs rarely last. Ground-breaking and stimulating designs may have a long life, but this is mainly due to their value as the first of a new breed, or out of curiosity.

New technology

New technology may be of great benefit; however, failure is always a constant threat and can bring a project down very rapidly and quite disastrously. This can lead to the complete failure for the whole building.

Flexibility

Designs that have little flexibility cannot adapt, and a slight variation in quite subtle issues may cause a catastrophic end to the building's life.

Avoiding fashion and its effects on the life of the building is impossible. However, the inclusion of some simple measures can make detrimental effects less likely.

- Look at how close the design is to mainstream thinking – are there elements or small principles that are outside the broadly accepted way of doing things?
- Use proven technologies – look at the provenance of any materials and products to ensure that they are robust, proven, and can be trusted to be durable.
- Built-in flexibility is probably the most beneficial attribute – the ability to change or modify is by far the best protection against future change.

Financial constraints in setting up a project

Fiscal models are also worth considering as they affect the building's viability, and are also partly responsible for the framework that sets controls at the project's inception:

- costs of running a facility
- write-downs

- costs of upgrading and keeping efficient
- costs of competition, performance legislation, changed expectations.

It is important to realise that at the start of a project, there is a 'to-and-fro' between the cost of the project and the performance – and, by implication, the specification and design. It is a matter of constant frustration that aspirational issues (primarily aesthetic) usually take second or even third place to the cost plan constraints.

This is not to say that we should not operate in the real world (cost and value are covered in Chapter 19). But it is important to put into perspective the overall benefits of less tangible elements to the project. Some notable examples do exist of the addition of value at the start of a project giving rise to substantial commercial success in the long term, confirming the initial design. One such example is the retail furniture company IKEA. Placing value for money at the top of the agenda, but in equal place with good design, has seen extraordinary success for the company as it has expanded to become a world-wide phenomenon.

We need to identify clearly the benefits of an approach that includes style, quality and design as equal partners with cost and value. Clear establishment of the core parameters is required, and is a strong driver for ensuring that the designed life value is the life value you get.

Chapter 14

Predicting the future

Practical whole life cost

So if whole life costs are not the answer, what is the practical and sensible way to proceed? One answer is to go about the whole process in a clear and evidence-based manner. The following is one way to achieve this, and an attempt to fend off all the influences that will cause a deviation from the prescribed route. All or part of this process may be followed, and the results will emerge in proportion to the inputs made and effort expended.

Setting out

At initial design, ensure the materials and processes match the brief. What kind of outcome is required? Is the establishment of whole life value of critical importance, of commercial benefit, an aspiration, or possibly a whim? It is necessary to be very clear about this. So many projects commit serious resources to a requirement that is just a 'nice to have'.

The essential questions in starting the process of whole life value are:

- What performance is required?
- What is the required life?
- What degree of maintenance will be performed?

The materials and elements involved need to be organised into the eventual package of requirements. The smallest of errors at the outset will result in large ones later in the building's life, and especially towards the end of the life cycle. Accuracy at the outset is therefore critical to the success of this process.

Having a clear idea of where you are going, and sticking to it, is essential. This is where comprehensive analysis is crucial. Quite often this is done implicitly, but key factors may be omitted from the thinking. It is critical that the full implications of any design are fully thought through. This can be difficult, given the trend towards shorter and shorter design periods. Getting the client on your side in this respect can therefore be half the battle. Clearly demonstrating the benefits of this approach will help.

Getting this authority can be a challenge – not every client needs, or indeed wants, this kind of analysis, in fact it may be counter to their business model. Many will want to produce a product and, in selling it on, will never consider the implications of its use and the impact it may have on the eventual users.

Clearly we have several economic currents that are counter to the principles of whole life value, and therefore to any element of sustainability. However, just as clearly, this state of affairs cannot continue, and more ethically charged clients are now embracing this process and turning it into a core business imperative.

It is therefore important that any analysis undertaken can stand up to scrutiny and deliver a worthwhile result.

Collecting the data

Data should be collected from a range of known sources. It is important to establish the track record of the material and its pedigree. At the heart of good specification writing is having a clear understanding of the match between requirements and performance. Normally, unless there are at least three independent sources of information, it cannot be relied on as being without bias. This is significant if the data are to be relied upon.

Sources of data will start with acknowledged references including:

- manufacturer, including historical data
- trade body
- testing laboratory
- comparison of similar manufacturers
- historical data from testing labs
- direct observation of examples.

Each to be defined and given goal setting status.

Making an informed choice

A balance of information, experience and interpretation is needed in order to arrive at a sensible conclusion that is relevant to the project, the brief and the circumstances.

How much tolerance is there?

Tolerance plays a large part in the success of a building. Some materials and systems have an inherently large tolerance; others hardly any. Understanding which you are dealing with is a big advantage when assessing the whole life costing of any part of a building.

There will always be a balance between initial cost versus the maintenance required for a given life. In principle, the higher the quality of the initial specification, the less maintenance will be required, all other issues being equal over the same period for lesser specifications. The challenge for the designer is to establish the critical issues determining the life and the required upkeep.

For two sets of specifications, from the same starting point, the outcomes will be substantially different depending on the quality and performance levels chosen. The care and attention needed for any given element of a building should be structured at regular intervals, taking account of the annual effects of weather and climate. The winter months will produce the most challenging effects for the exterior; the summer, with long periods of radiation, thunderstorms and flash floods, also causes wear and tear, but normally interspaced with calmer periods. It is during these periods that inspection and maintenance can take place. Interiors also need periods of cleaning, inspection and maintenance, and summer is a good time for extensive operations to be carried out here as well.

Maintenance intervals

For these reasons, significant maintenance should be scheduled as at least an annual event. Possibly the best arrangement is for checks to be annual; 6-monthly; 3-monthly; monthly; with cleaning undertaken at least every 3 months.

By setting up and ensuring regular cleaning, any maintenance issues can be identified. If these are picked up early, the work will always be less than if left. Building management teams need to recognise the necessary processes and ensure they are undertaken.

Many manufacturers set out in their literature, and as a requirement of their guarantees, that regular cleaning inspections and potentially maintenance are carried out at not more than 3-monthly intervals. If this is the case then, barring major accidents, there is every possibility that the building will perform as designed for the full length of its life.

However, it is clear that lower-quality specifications will require more effort, and more care and attention, to maintain the same level of performance. As the building ages, this element will also increase. Looking at the original specification to avoid this effect will necessitate greater capital expenditure. But overall, the resources used will be less.

We can consider the resources used as an initial significant amount to generate the product in the first place during manufacture. Fashioning it and installing in the building will require further input. Entropy begins after completion, due to the ravages of environment and use (see Chapter 21). Maintenance and cleaning can be considered as counteracting this entropy.

Ensuring an understanding of these principles and explaining them to the client at the early stages of the project are essential if a rational approach to whole life matters is to be undertaken. Establishing the whole life cost, in both fiscal and resource terms, is the next stage. It is possible to identify the cost quality parameters, each having a required cleaning, inspection and maintenance pattern. This can be the cost and resources allocated for the total life of the element.

Undertaking this for all elements of the building will be possible one day, when total elemental design, procurement and construction is common. However, currently I would suggest this is practical only for the significant elements of the building: the structure and foundations, cladding, roof, and the main elements of plant. Nonetheless, the ability to analyse these at several quality levels and to project their needs over their lifetimes is a powerful tool to identify the correct solution.

Spelling out the implications of change

The design and (hopefully) construction teams will leave the building in the hands of others at completion. This is a crucial point. Has the right information been transmitted to the right people, so that clear understanding is achieved? This is another danger point, where misinterpretation can undo all the hard work that went into establishing and creating the building and launching it on its life in use.

Table 14.1 The process map

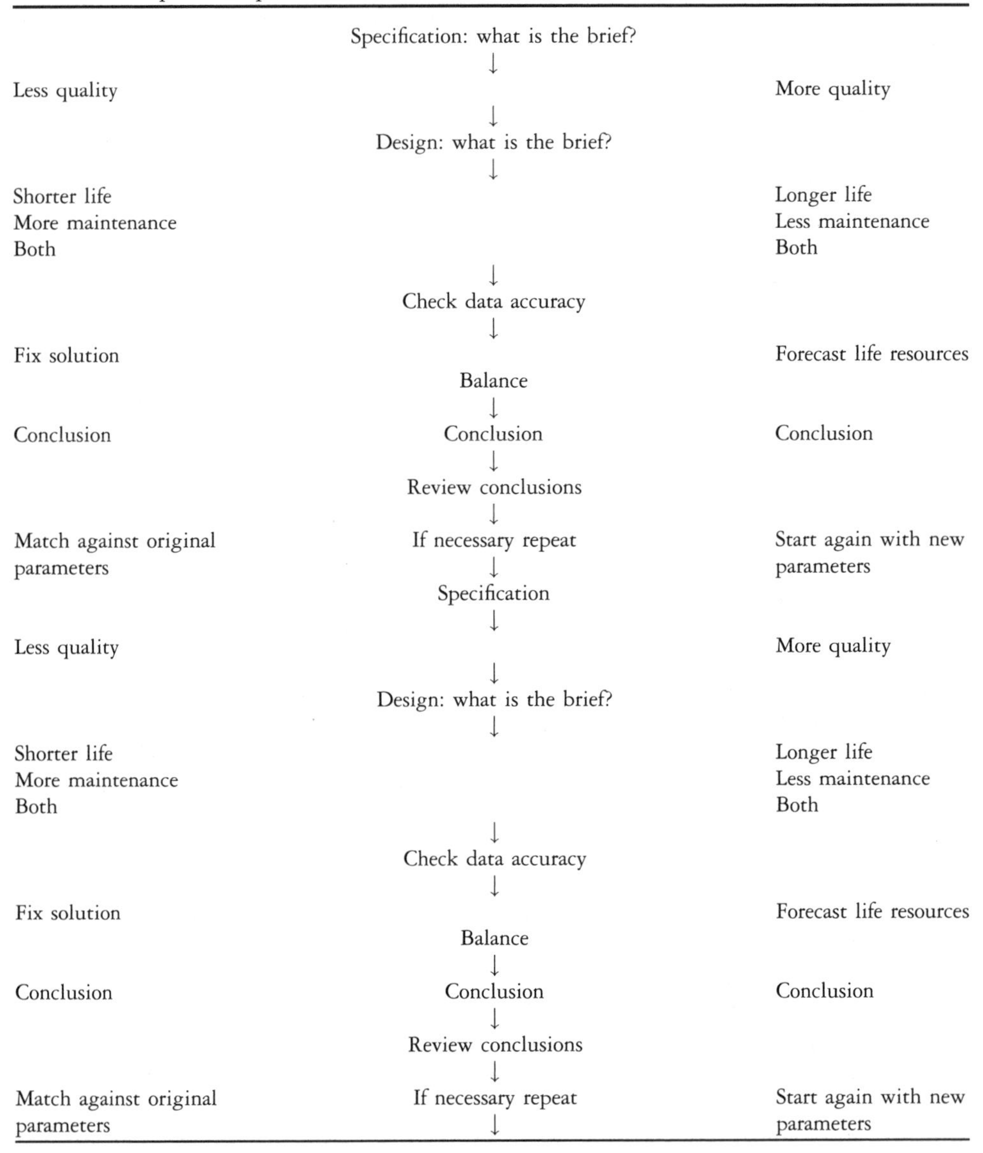

	Specification: what is the brief?	
	↓	
Less quality		More quality
	↓	
	Design: what is the brief?	
	↓	
Shorter life More maintenance Both		Longer life Less maintenance Both
	↓	
	Check data accuracy	
	↓	
Fix solution		Forecast life resources
	Balance	
	↓	
Conclusion	Conclusion	Conclusion
	↓	
	Review conclusions	
	↓	
Match against original parameters	If necessary repeat	Start again with new parameters
	↓	
	Specification	
	↓	
Less quality		More quality
	↓	
	Design: what is the brief?	
	↓	
Shorter life More maintenance Both		Longer life Less maintenance Both
	↓	
	Check data accuracy	
	↓	
Fix solution		Forecast life resources
	Balance	
	↓	
Conclusion	Conclusion	Conclusion
	↓	
	Review conclusions	
	↓	
Match against original parameters	If necessary repeat	Start again with new parameters
	↓	

It has to be used, cleaned, maintained and cared for in the manner prescribed in the designers' minds. Only then will the carefully crafted outcomes be realised.

One example is the building manager who happened to observe the roof drainage system in a storm. He saw the gutters backing up dangerously, and made up his mind to do something. Returning to the roof in good weather, he was convinced that the builders had wrongly left protection in place over the downpipes, and removed the plates fixed across the outlets. In the meantime, the service company contracted to ensure the building was well maintained replaced the plates, knowing them to be critical to the operation of the system (which was based on a siphonic design). This cycle of removal and replacement went on for several months before both parties, by chance, found out who was doing what.

A simple, clear briefing at handover could have resolved this at a stroke. Many buildings suffer from this complete detachment of the design construction team from the facility management team, with dire consequences.

It is clear that, if whole life cost is to have any benefit, there needs to be joined-up thinking, and principles need to be carried through without modification. To ensure this is the case, designers need to ensure issues are embedded to ensure the building's future.

Sustainable characteristics

One of the benefits of a structured life-value analysis is in determining the sustainable characteristics for any development. The ability of a building to be of benefit as it comes to the end of its design life will depend on a number of criteria.

- Is it still viable under its current use? This will be based on commercial issues, fashion, and external influences such as location – how much the site is worth can be more relevant than any other consideration.
- Repair may be a first consideration. Making the building fit for a new life cycle is a very good use of materials and resources. This can be seen with current buildings built many decades ago – careful refurbishment not only brings new life, but also adds character that would not necessarily be there with an entirely new building.
- Reuse of the building may be a viable option. At the end of a design's life, there will almost always be a further life cycle that may well start a new whole life judgement. This will put a different conclusion on the original scenario and change the values that had been set. However, due to the range of unknowns and issues that can affect the building's life, this cannot be seen as anything other than a natural course of events.
- Recycling is another possible outcome at the end of the design life. Again, it is important not to make too many assumptions.

Systems and fittings from the building will be of some value, and therefore could be recycled. Materials may also be reconstituted and reused – such as aluminium, concrete and steel. The commodity value of these items at the time will determine just how much, and in what way, they are recycled.

Chapter 15

The ten-point strategic plan

How can we cut through the problems? A clear plan is needed that can resist the pressures of the normal turn of events, hopefully delivering something tangible. This requires clear thinking, a committed client, and a team that can see and embrace the benefits.

Ten practical steps to generate and organise your whole life value programme

In order to make this work, I suggest here a ten-point plan that can be written into the project procedures. The most important issue is that this requires a long-term commitment. If this cannot be guaranteed, then there is little point in starting.

Objective 1 – confirm the brief, get the client on side, and clearly establish the requirements

This means that everyone in the team understands what is required and that the objectives in relation to the brief are well established. While there may be gaps in understanding, or about how this will ultimately be achieved, desired outcomes are clearly defined. What benefit can life value bring to each of the elements of the brief? It is important to spell out exactly what this will mean to the project. This should establish an additional baseline for the project.

Objective 2 – ensure the whole team works to this requirement throughout the design, procurement and construction process

As the project progresses, the make-up of the team will change – especially in the later stages, and almost always at handover. Commitment to the programme and principle will therefore have to be re-presented, restated and made relevant at regular intervals throughout the project's life.

This is not only about revisiting the message; it is also about ensuring the process of revisiting is undertaken. The whole team should have a clear commitment to the principles of whole life values, and continue to hold that belief.

Objective 3 – agree mechanisms for checking accuracy, and actions where deviations from the requirements are identified

One of the most serious factors preventing implementation of good whole life values is the changes made at every level of the project. These can chip away at the initial quality and direction of the building and destroy its longer-term performance. This can be part of the quality assurance objectives (if used) or part of the established procedures. The project should have procedures for control, feedback analysis and implementation of any changes required.

Objective 4 – consider specific uses for every element of the design

As the design progresses from concept through outline to detailed design, the subject is increasingly narrowed in scale. In other words, at the start of a design the concept issues can only be looked at in general, while at more detailed design stages it is possible to consider the totality of the building's components down to the smallest fixing or the finest level of finish. Throughout, the same principle of attention to detail is required – what is its use? What is the required life? Is it fit for purpose?

Objective 5 – ensure consensus over the implications of the whole life values and costings

Whole life values and costings cannot be established in isolation. When one part of the team makes decisions unilaterally, they are never as successful as if the conclusions are arrived at by all. Many will assume that those in authority own and control this process; however, in many cases lack of engagement or buy-in can result in problems.

Potential interactive effects may include the following.

- Size – component size affects manufacturing tolerances, placement tolerances and replacements.
- Weight – very heavy items are difficult to transport, locate and install. Similarly, they are difficult and expensive to replace.
- Density – very dense components are again difficult to install or replace, although due to their obvious robustness they may exhibit very good life characteristics.
- Material type – a vast range of materials are available to today's designers, and their characteristics must be considered. Using a material in sympathy with its nature is a very important measure to ensure its maximum potential life.
- Colour – the colour of a material often gives a very good clue to its make-up, chemical composition and, ultimately, ability to withstand the rigours of its life in service. Bright colours, dark colours, and the red part of the spectrum are poor performers.
- Resistance to environmental effects should be understood in this context. Several are related to the site and location of the building. Key factors include:
    - moisture – rain, humidity, steam, snow, ice
    - radiation – sunlight
    - temperature – excesses of cold or heat and rapid changes in between
    - movement caused by wind temperature, pressure or geological movement
    - contamination – pollution including particulates; chemicals in solid, liquid or gaseous form.

Objective 6 – ensure there is a baton-passing process in place

This is important for confidence that all the work carried out to ensure life objectives are met will be of benefit to those in authority at the later stages of the building's life.

Objective 7 – clearly record decisions and details stemming from the original design

For example, the cladding will be specified to a certain life – this will mean a minimum standard of product assembly, workmanship and quality at handover. Following handover, this life will be certain only if maintenance, cleaning and repair are completed to the stated standard.

Objective 8 – ensure the data is rigorously checked

False assumptions and incorrect information will mean the system fails.

Objective 9 – ensure post-completion actions are followed up

It is important that post-completion actions are written into agreements and adhered to, and that there are checks in place.

Objective 10 – agree milestone and check points

These should be at all significant points in the project.

Concept design

This includes key parameters, life criteria and implementation processes.

Detailed design

For every element, it needs to be established how it will be embodied in the final specification. Are there any gaps or overlaps?

Planning consent

Are there any constraints or changes caused by planning conditions?

Tender acceptance

Has anything in the tender changed the established criteria? What is in the small print? Is the required level of detail present, is the quality maintained, and are the processes robust?

Start on site

Check that nothing has changed, information is correct, and change-control processes are in place.

Letting of all major packages

This is another check of the details and assumptions; check that changes in one package are reflected in others.

Foundations completion

Have all criteria been met? Have any changes been accepted? If so, do they alter any other elements? Were any unknown issues uncovered? How were they controlled and resolved?

Structure completion and topping out

Have the criteria been adhered to? Have any changes been made – especially interfaces of movement joints, connections, fire protection – that will have knock-on effects?

Completion of water-tight and air-light fabrics

This is a very significant point in the project. Have all the junctions been completed as designed? Have any changes been made? Have all areas been tested and verified as correct? Often too little time is spent on this element.

Completion of services installation

Have all services been installed and tested as correct?

Completion of interior

Have all interior fittings and fixtures been installed as specified?

Practical completion

This is a final check prior to handover, occupation and start of use. Are all elements of the building as robust as possible? Have they been proven to be as originally designed and specified? What measure of confidence is there that the life values established have been preserved in the final build?

The 'baton-passing' process

Do the client, tenants and occupants understand the build? Is it clear what maintenance and servicing the building needs?

Start of occupation

This point is often the start of more changes – have these been coordinated with the building's original design? Are there any conflicts that will cause problems or devalue the robustness of the building?

First anniversary of occupation

This point is significant as it often brings the end of defects liabilities – and is the chance for the contractor to remedy any outstanding issues that have arisen. It is normally clear at this point if there are going to be any long-term issues that will ultimately affect the building's performance.

And finally

This may appear to be just good practice – but even if there is no direct, tangible evidence that such measures are needed, this process is still very much worth the effort, and real benefits will be achieved.

Chapter 16

Key players

Consultants

To many consultants, whole life costing is an aspiration, but not yet a reality. High-profile, larger projects attract a lot of interest in this area and have the time and resources to devote to the 'established systems' – by which I mean that a whole life approach is expected to be included. This is understandable, given that much of the industry functions on the basis of adopted common practice, although many in the team may have doubts that it will be undertaken.

On one hand, there are also many projects with neither the resources for, nor the expectation of, a consideration of whole life issues. This creates – or at least contributes to – a two-tier system: large and small. On the other hand, projects fully embrace the concept and take on the process of endeavouring to establish the project's whole life cost; on the other hand, the subject does not get off the ground and is currently not part of the agenda.

In my opinion, both camps are wrong and there is a need for a more open-minded approach. Consultants are an unusual breed, sitting between the client and the 'hands-on' construction world. They can be both immensely influential and significantly destructive, sometimes at the same time. We, as a society, look aspirationally to consultants. Most people know them as professionals providing the moral high ground, the innovation and the ethical correctness required by the built environment.

Pressure caused by ever-increasing demands on resources, and limits on energy and carbon, increase this dependency. We look to consultants to ensure the 'right' thing is done. But their influence, logic and ability to argue the case for the good struggle to carry any weight against the hard steel of commercial viability and political expediency. Consultants have, above all, a duty of care to their clients, combined with the principles of their professional standing. This implicitly means delivering the best possible outcome, given the criteria and resources available. It also may mean that they challenge the client's approach and attempt to add value to the enterprise. However, this is only rarely successful; often key issues are overlooked, or developed in a way that is highly questionable if our goal is the best possible outcome. The bigger the project, the bigger the effect.

This also assumes that consultants have the appropriate level of understanding for the process they are recommending, and are experienced with it on live projects. Often, there is little understanding or experience enabling a full and reasonable exchange of views and allowing sensible progress to be made.

Large, high-profile projects tend to have a whole life costing programme applied to them, often backed by a considerable amount of public relations. For these projects, this may also provide a lifeline to back up the design rationale and ensure that accountability – public or otherwise – is satisfied. This can be tuned to justify all manner of decision-making and, in some circumstances, to explain the unexplainable.

However, this can be merely window dressing, disguising the hard facts and wrapping them up in a gloss that will fool most people for many years until it is too late – until it is quite clear that the analysis, the clear decision-making and good intentions are as rotten as the prematurely ageing facades of the buildings to which they are applied.

Consultants are likely to earn fees from the life analysis they undertake, or may have used this expertise as a 'commercial edge' to get appointed to the project. This demonstrates how hard the appointment process is, and how competitive and potentially flawed the selection can be.

How do consultants support whole life costing or value analysis?

Consultants are keen that their clients see them in the best light. Generally, this means that they have a range of skills that can be deployed on any project. The client's overall respect for the consultant is centred in the belief that the consultant can employ these skills to ensure success, in whatever way the project needs. Making clear what they can do and adding value, as well as looking after the project's healthy outcome, are essential factors. In addition to the basic tasks they do to earn their fee, all consultants are keen to add additional benefits and to offer a picture of rounded experience. This is often where whole life costing, not a mainstream activity, may be held up as an additional benefit of employing a particular consultant.

It is easy for consultants to follow a policy line and avoid rocking the boat, which makes positioning of whole life issues within the project quite difficult. The balance between principle and remaining employed often arises, and this is not the only instance where the consultant may feel its better to keep quiet and go along with the *status quo* rather than ploughing the efficiency furrow.

In a rapidly changing world, however, the future for consultants may see a different approach, as new issues are rearranging the drivers and pressures. Sustainability, zero carbon, low carbon, social responsibility, and health and safety are all drivers that are hard to ignore, and need a realistic and well-considered response. These will change the way we think, use and define whole life values across every project, and issues that arise from them will, I hope, at last give rise to logical and appropriate consideration of how we plan and act. It is crucial that we understand and manage the factors that arise. Lessons from the past seem to suggest that when clients identify issues as part of their agenda, the team pays attention. If consultants do not rise to this challenge, then others will emerge to undertake the role. We have seen this with the proliferation of specialists over the past 20 years, moving in on and marginalising formerly traditional roles.

Some large consultancies have championed whole life cost and value. They have an interest in the cause, and rightly so; and they have the resources and the profile to advance in areas where progress is slow and leadership is needed. These models are interesting and may be of use in some areas, but largely run out of steam during completion or once the building is complete and occupied.

Consultants in future will need to adapt and respond to these challenges, offering clients robust answers. Equally, there is a need for the rest of the industry to understand what they need to do and to improve on the performance achieved to date.

Building users and facilities managers

After the designer, consultants, specialists and contractors have gone, what happens to the building's needs and requirements? Why do they become invisible, and why is it largely assumed that a building will continue perpetually without due care and attention?

The most serious impact of the whole life costing issue comes to rest on two groups: clients and users. But neither is especially engaged with the process of building procurement, and generally they seem to be oblivious to these issues. Part of the problem is they have different perspectives, and regard it as 'not their problem'. They may be viewed as either passive or active participants. Passive users are those who use the building, as a place of work, as a dwelling, or for some other purpose. Active users are those who have to maintain and clean, or have a responsibility or interest in maintaining and cleaning, the building. Passive users rely on the building and trust that the basic performance criteria will be delivered, ensuring the required functions, protecting them from the extremes of climate, and providing a safe, secure environment. But how they use the building, and whether they treat it with respect, will contribute to its long-term performance. Equally, the performance of the building will affect their comfort, wellbeing and safety if it falls below the clear norms of basic practice. The extreme example here is a building's performance in an earthquake.

In particular, the parameters that need care by both building users and operators are:

- thermal performance and responses
- weatherproofing and ability to shelter the occupants and resist weather extremes
- performance of fabric (in addition to the above) for security, durability, ease of cleaning and maintenance
- performance of systems, heating, cooling, ventilation, communications, security
- health and safety issues, especially slips and trips, hazards and environmental control
- microbes and ventilation
- cleaning and cleaning products.

None of these issues is complex, and none is difficult to resolve.

Engaged users will have the task of ensuring the building stays clean and performs to an optimum level over years of use. There is a strong correlation between the effectiveness of the original specification and building quality and the level of difficulty for this group. A correctly identified specification, designed and built correctly, will be far easier to clean and maintain than a relatively low-specification, poor quality building.

It is also very common for designers not to seek the expertise of the cleaning and maintenance industry, which is an obvious blind spot. Clearly, the cost of building cleaning and upkeep is not seriously considered in most projects, even when whole life cost is taken into consideration. To me, this is the great conundrum: why, when so much effort and resource is put into establishing a so-called whole life analysis, is this thrown away at completion and ignored during probably the whole of its occupied life?

I suggest that the answer is found in the way we fund and set up construction projects. There is a series of interested parties who, their part complete, will hand on the project. At each stage, value is added, but equally detailed knowledge is lost, in particular understanding of the more embedded values in the building. This is exacerbated because maintenance and cleaning is normally segregated, placed in a separate budget and never given a very high profile.

There is also a strong feeling that this remains an issue that is hidden from most of the building team, and certainly from the building's users. The accepted principle, led by senior management, is that the building will not need anything but the most basic of maintenance and care. This translates to a reliance on basic measures until something breaks, then an attitude of 'just fix it and carry on' prevails.

This is not sensible in any circumstance, but if the original building is of poor quality, this will be amplified in use and issues will follow: leaking roofs, draughts, doors and windows that do not close correctly, heating or plumbing leaking. Many issues occur that are unseen until it is too late; or that result in knock-on failures before they are addressed. With this, the cost increases, and the potential for the building to go on to achieve its designed life is significantly reduced.

There is something of disaster theory in this process. Issues that are not considered important by management until a serious failure occurs may then be given time and a budget by senior management. So waiting until a serious failure occurs will result in corrective action. This is not good business, nor is it sustainable. Corrective measures will cost more and take more resources than if the damage had been prevented in the first place. No failure at all would be the best possible situation – which means buildings that are specified to last and to be maintained for optimum performance.

Often this whole process is too difficult to manage, and the effects are largely unseen, or are not regarded as a serious enough issue. Greater emphasis on, and understanding of, resources used, waste and energy demanded will increasingly result in problems for the business unless the root causes are regarded as serious.

Legislation, shareholder pressure and business efficiency will all have an impact, as demonstrated by the growth in membership of the UK Green Building Council over the past five years. The agenda is beginning to change.

The following examines some examples of poor management that I have come across, and how this directly affects the outcome from a whole life perspective.

Cleaning

We all know that the quality of cleaning can have a substantial effect on the look and feel of a building and, incidentally, on the image of the company that occupies it. Cleaning can also affect the life of the finishes and exposed materials and often their life beyond.

Can the cleaning that is applied get to all the areas needed, into the corners and onto all the surfaces needed? Are the right materials used to clean? Does this effort cover all the necessary areas, inside and out, including the often-ignored roof? Can difficult-to-access areas inside voids, pipes, ducts and chambers be cleaned correctly? If not, the building will degrade and age earlier than designed. It is of great concern that maintenance difficulties contribute greatly to buildings' curtailed life. Designers can help by making easier access part of the concept, although often this a target for value engineering.

The following considers specific areas of a building.

Trafficked areas – floors, entrances and circulation spaces

These are subject to quite obvious tracking patterns. The main areas of wear and the inevitable matting, added to the centre line of the space, betray a building under stress. The entrance is the first experience many have of a building – often a poor one, due to the chaotic nature of the finishes, or the fending off of problems with the finishes. This is extremely common, even in recently completed buildings, and unless it has a serious impact users hardly notice it.

Lifts

Lifts have two inherent problems: the quality of the car; and the reliability of the cars to go where they are instructed reliably. Both require constant attention. In many buildings where lifts are heavily used, car finishes can age rapidly, making the appearance look poor. Attention to quality is essential – not just for tired-looking floors, doors and frames, but controls and signs also age rapidly.

Equally, the lifts' ability to deliver a reliable performance is central, and most will understand the need for regular, planned maintenance. While safety is ensured by a clear obligation to maintain all relevant systems, other areas are often ignored.

Ironmongery

Ironmongery is described by some as a building's primary interface – it is the element that most users touch directly. The look and feel of the ironmongery affects the users' personal interaction with the building, and they will form an opinion of it from this experience. Because of this, it also wears heavily and needs to be cleaned and maintained continuously. Ironmongery also owes a lot to the original specification, and many designers perhaps do not spend enough time considering this aspect.

Roofs

Most users will not have detailed experience of the building's roof, but the maintenance team will – or at least they should. Despite being the most important element in a building, roofs generally do not receive the attention they deserve. Particular issues include ponding and leaking of either the roof or the drainage system. Overflowing of the drainage system and damage due to weather, heavy-handed modifications or debris all play a part. Thermal movement and solar radiation, opening up joints to attack by ice and rain, are equally aggressive.

Roofs have to cope with all these issues and provide protection from the elements. But at all stages of design and construction, corners are cut and principles diluted, and most result in a compromise. Across the built environment, roofs are largely left unmaintained until something breaks.

Heating, cooling, ventilation and associated systems

The performance of heating systems has come into focus as the pressure to make systems more efficient causes the tolerances that once were built into the design to be slimmed down or removed. Heating systems now are designed to be just adequate, and rely on the improved specifications of fabric and ventilation systems to ensure occupants achieve comfort.

However, while this may be the theory, often the combined performance does not stack up, and the simple ability to keep warm is inhibited or lost. This may be due to 'value engineering', poor workmanship, or simply poor design – but whatever the cause, its effect on the predicted whole life cost of the system, and even of the building, is considerable. Poor performance of such a simple requirement will devalue the building in the eyes of its users and occupants. Some may have the system modified, and some will struggle on, maybe adding additional heat sources or just putting up with the prevailing circumstances. In either case, the whole life values and costs are now in tatters, and any thought of matching the longer-term goals is completely lost.

Cooling and ventilation can be seen as mirroring heating, although faults are much more common. This is because modern cooling has not been with us that long and the margin for error seems to be greater, or perhaps the expertise is just not available. Cooling systems seem to be much more temperamental and require more than their share of attention. Again, a poorly working system will make any long-term assessment almost meaningless and will greatly devalue any forecasts of performance and replacement targets.

Windows and doors

Windows and doors are the most vulnerable element of building fabric, but air seal is critical to everyday and long-term performance. With energy input under scrutiny, any uncontrolled leaks must be at a minimum. This is still a relatively young area for construction in the UK. Focusing on ensuring the fabric delivers a constant, high-quality seal is still very new in some sectors of the industry. Performance data are still relatively rare, but the data we do have indicate that performance is poor to patchy, which gives a clue to the prospects for delivering better performance in future.

Any fall-off in performance will affect the outcome of the whole life model. This is complex because there are so many small areas, junctions and materials that can all contribute to 'failure'. Additionally, this failure can be manifested in so many ways as to make any proactive prediction method almost impossible. Proven durable and robust solutions are hard to find. It may be that the only way will be to ignore the majority of potential errors and accept that an overall life value is more important than the small, detailed errors in all the components.

Windows and doors are probably the most important element here. How they seal over many years of service will come down to the technical design and the durability of the materials used. It is possible to specify products that will perform for many years and with little decrease in performance, but the tangible differences between those units and other elements that will fail early are not obvious and require a high degree of understanding at the specification and, critically, the construction stages. Often these factors are ignored or not understood. Therefore early failure is built into the design, and no amount of analysis or paperwork will change this outcome.

Fixings and fittings

Corrosion of fixings is often never considered, especially where they are at the junction between components manufactured by different companies. Often unseen, the combination of different materials can result in chemical action and accelerated ageing. These will never deliver the service life that was expected. Bimetallic action can add to the chemical effects and accelerate the destructive action, increasing the problem.

Glues and sealants are increasingly used in significant assemblies, which takes us into unknown territory in terms of service life and time to failure. Newer methods of plumbing composite materials rely on correct materials science and accurate assembly, and cannot be checked on site.

Clients

Clients have a very strict view of the world – their main concern tends to be with the overall success of the project. They may focus on a number of issues, but may not see them in same way as the designer. This can cause difficulties, especially as regards whole life costing.

Whole life costing can be a euphemism, a real objective, a marketing tool, or seen as part of several other objectives. However, if tabled it will be part of the client's business plan – and the designer forgets this at their peril.

Understanding what is in the client's mind should be paramount to both designer and consultant – regarding their business, what exactly they think they are engaging you for, what you can do for them, and how it will all work out. These objectives, more often than not, will show divergence or even complete opposition. So how do we square the circle?

Respect

Respect is at the centre of most relationships in business, but it is an intangible and difficult attribute to define or to cultivate. The need for respect in the relationship between client and consultant is paramount. Many project difficulties can be explained by a lack of respect between the key team members. Respect must be earned – it is not something that is freely given and, once attained, it can be easily lost. I have been in many meetings where respect is lost and gained, almost in the twinkle of an eye.

This is especially the case when working on issues that need clients and the team to 'buy in' to the ideas and principles being suggested, and never more so than when taking a whole life costing approach to the project.

Respect is not a factor that is easily come by or definable. It is the tacit approval that team members may have for the actions and opinions of others. It is crucial, however, in determining the project pathway and the elements that will come together to drive success. On many occasions, issues that would otherwise cause problems can be overcome when individuals have respect for each other. The reverse is also true, and this is at the heart of many situations where problem after problem seems to surface.

Confidence

It is important to have confidence in the process – what it can do, what it can't, and possible outcomes. Being confident and talking clearly in jargon-less, calm English is always a good place to start. Including direct experience and project examples is also beneficial – but avoiding any directly negative themes.

Business plan

Demonstrate that you understand the client's business plan and how this might support it or fit in with it. While the timeframes may be vastly different, it should be possible to present a case that helps the business plan, and this will be all for the good of the client.

Changing landscape

It is essential to be aware that the industry landscape is changing. The issues we confront today are different from those of only a couple of years ago. They change our perspective, our approach, and the answers we can deliver. We need to be aware of these issues and ensure they are factored into the approach taken. To some degree, this has already happened in much of the whole life arena.

Objectives

Clarity of objectives must be defined, especially the whole life objectives and how they are going to be delivered, not just at procurement and construction, but over the whole of the building's useful life. Identifying objectives may seem obvious, but it is surprising how often these are not explicitly stated or spelt out in the brief so that the whole team is aware and clearly oriented in the same direction. A significant problem with whole life issues is diffusion of the initial energy and enthusiasm as the project progresses until there remains often only the hint of the ghost that is the whole life analysis.

Sustainability

Part of the moral imperative implicit in the principles of sustainable development is the idea that the resources used will maximise the energy and materials available – always striving for the longest possible life cycle. It is therefore a cornerstone of sustainability that life-cycle analysis will be part of the design, and whole life value a major consideration.

Owner–occupiers

The owner–occupier is always more interested in the performance of their asset than either the tenant or landlord, for good reason – they need to get as much out of it as they can. So the market leaders, or clients to whom the whole life cost and value will have most relevance, are owner–occupier clients.

Outside influences

Often we can see the effects on contractors of outside influences. This may be directly, as in a new law; or indirectly, as in an economic change.

Business drivers, simple economics, or more national influences such as legislation or a shift in politics have a significant impact on what is achievable or desirable. It is important to understand that options are limited, and often the opportunity and drivers are simply not there. Often the safest and easiest route means the usual formula or process is just repeated. That is not an excuse, however – there are many complementary issues that can be changed as part of the usual process, with an outcome for the better.

Having taken all the significant issues into account, there is still a need to consider unforeseen effects or knock-on consequences, which occur with startling regularity. Such is the pressure to deliver that, while good intentions will prevail at the bidding and early stages, by the time a project gets to site, most of this will have evaporated. On the most highly charged and high-profile projects, only a thread of whole life costing will get through.

Whole life costing is used by contractors in a number of ways that are quite interesting to examine; however, these tend to have little effect on projects in real terms.

Getting the job

Everyone in the construction industry understands that it is a competitive environment. Such an animal can be difficult to feed sometimes. If contractors are to survive, they need to win projects; and in order to win projects, they need opportunity, experience and luck. Sometimes good judgement will also enter the equation.

Having found themselves in a bidding situation, many contractors will attempt to ensure they have all the bases covered. Bidding these days involves more than pricing the prelims, day rate and insurances, followed by a 'finger in the air'. It is essential to give the client a targeted presentation or document to have any chance of success. Factors such as sustainability, corporate social responsibility, and health and safety figure highly in the rankings. So does whole life costing.

On larger projects, which may attract sophisticated bidding teams, a significant resource will be devoted just to this element; on smaller projects, it has to take its place in the food chain. Methodologies will vary: for some, complex systems and costing will be applied, targeting the prime building elements such as roof cladding, structure and services. The important thing is to convince the client that the team is worth backing and has the experience and focus to deliver the project as required.

Once appointed, the job starts in earnest and the inevitable reality kicks in. Detailed design compliance with the brief regulations and the inevitable actual costs start to build pressure. It is clear, from my experience, that the majority of cost plans are based on generics, and from the outset set a sequence of events running that always lead to cost-cutting and a reduction of standards throughout the project to completion.

The application of real costs to the realities of the detailing leads to the process of 'value engineering', which will see most of the value-added elements, including ethical and appropriate sourcing, reduced or eliminated. Reviewing the project on each cycle will result in stripping out of any but the essential components. This will have a substantial effect on the potential life of many, if not all, of the building's elements.

When we get to completion on site, very little of the whole life value will exist, and much of the initial enthusiasm will have long been forgotten. It is greatly disappointing that this is usually accepted by the client. For the most part, clients are philosophical about it – and in any event will be passing the building on to others. Other departments, or tenants, or occupiers will take on the building, and will have no interest in whole life issues. Their responsibilities take over, and using the building as part of their business is much more important to them.

Critical to this situation is the importance placed on whole life values by whoever is in authority over the building, whether directly or as an investment. Normally this is very low on their scale of importance, and consequently any failures will not be flagged up.

Chapter 17

Client experiences

Clients are either unaware of, or insufficiently impressed by, whole life costing to consider it worth including in most projects. Many projects I have been involved with have debated it at the start, but considered it irrelevant. This may be due primarily to the financial structure behind most projects in the commercial world. It may also be that whole life costing is not considered critical or relevant.

Most construction in the UK involves funding from third parties and to a large extent projects, once complete, are passed on, with many enterprises taking their cut from the profit made at the exchange of ownership. It remains to be seen, following the recent financial turmoil, if this state of affairs will continue in the same way as in the past. The fight between value and performance is all too apparent in the client's dilemma.

Projects will, of course, have a design life. This is more a gauge of quality at the build completion than a real assessment of the 'life' of the building. In the briefing and design development stages, clients will often set a design life. This enables the team to set the specification standard and to provide evidence to the funding organisations that the project will have financial backbone. One possible exception in this is the government client (see below).

Having set some parameters, requirements are usually very general: structure, façade, windows, finishes, services, for example. Client's requirements will rarely go deep, and are often included for the sake of the usual process, rather than for any considered tangible benefit.

I have asked many times for the definition of design life in these circumstances. Some quote the British Standard; some quote several reference documents that have been well used over the years; and some will provide a relatively arbitrary definition based on the feeling during the meeting at the time. All of this points to a relatively devolved industry that does not have any real feeling for whole life issues, which rarely have an impact on early decision-making.

Rarely, if clients are not motivated to take up an issue, will these factors be introduced into projects. The client is the driver, and the team usually respects the client's position – if the client doesn't pursue whole life costing, then it won't be part of the project's make-up.

But many clients are waking up to the idea of sustainability. This is being driven by increasing interest in all things environmental as a result of climate change issues. Many clients now have a corporate and social responsibility programme, and many will also have shareholders pressing this home. This leads to much more consideration of how we use materials and resources, and the effects on people and the environment.

This is by no means universal, but it is significant, is having a direct impact on the way clients approach projects, and may in turn have an impact on whole life cost.

Ten years ago, assessments using the Building Research Establishment's Environmental Assessment Method (BREEAM) were regarded as an interesting exercise and applied to a few projects. Five years ago, around 30% of projects required not only assessment, but also minimum acceptable levels. Currently, nearly all projects make a BREEAM assessment mandatory, with most demanding a better than average pass. This has been raising the effect of quality in environmental areas, and a knock-on effect is that the inherent life of many building elements has risen as a result.

However, this is despite whole life costing, not because of it. Interestingly, some now challenge the value of the BREEAM process itself in delivering proven benefits. Others look to rival systems. This may have the effect of bringing whole life issues out from the shadows.

The one exception to this has been government projects. This is because government as a client has taken the view that it is procuring on an owner–occupier basis, and therefore requires best value for money. The odd man out here has been the private finance initiative (PFI). I certainly consider the PFI side of the industry will show significant early failure or increased maintenance costs over those forecast, due to the procurement process employed. But in general, government has championed whole life costing issues at all levels of projects in all sectors. It remains to be seen if this effort and commitment will be borne out in tangible benefits.

Most government work has to be formularised. Accountability is all-important, and is an obvious method of ensuring that a project demonstrates the required attributes. However, I question whether this process delivers value in its own right, or if any tangible benefits will result over time. The analysis carried out allows for a paper response to the issue of quality; but not accepting quality in its own right shows a lack of confidence in the procurement and construction process. It is this disconnect that is the problem. Arguably, this is in contrast to the Victorians, who did not lack conviction in this area.

The wall of bureaucracy within PFI also ensures that if any questions are asked, there is a paper chain that can be pulled out to justify the decisions made. But did these help with the Scottish Parliament? I think not. That is an example of a project with some of the highest levels of quality available, yet there have still been well publicised failures just a few years into its life. This will be a fascinating building to monitor over the coming decades to see if the theoretical assessment bears any resemblance to the reality.

My aim here is not merely to criticise government procurement methods, but rather to draw out the factors behind the use – or not – of whole life costing, and any tangible benefits that may result. We need a more obvious and robust way of accounting for the life of materials and assemblies for clients to use as part of their project procedures.

It is clear that there is a proportional relationship between quality, life and resources. I have seen choices made at the design stage that are clearly wrong, but that were made on the basis of the initial cost plan, irrespective of any other considerations. This is effectively a waste – of the materials used; of the energy and effort used; and, above all, of opportunity. Every new building project should be seen as an opportunity to produce a new asset, not just for the client, but for society.

The essence of sustainability is that resources and opportunities are used wisely; in many cases, currently, they are not. This sad reality is due to the prevailing circumstances, and there is little hope of change in the near future.

How does this work in a real-life situation?

Some years ago, I entered into a heated debate concerning roofing. Roofs take considerable punishment and need to work at 100%. Most clients become extremely upset if there is a roof problem. While many problems that occur with buildings can be an irritation, and will be fixed eventually, any problem with a roof usually results in a call for the cavalry, and in most cases their support troops as well. It is therefore understandable that debates on roofing specifications, details and designs sometimes reach staggering proportions.

This story relates to one such occasion, involving a large, commercially funded, ultimately government job. We were faced with a relatively conventional requirement: a flat roof covering a considerable area, with no special features or distinguishing issues. The client, however, was very strict in their requirement for a life of 40 years. This was coupled with an overall requirement on the project that whole life values would be analysed and the best, most economic route chosen for all significant elements. As always, the devil is in the detail, and it took considerable time for the team to even begin to discuss the actual building materials, as those involved debated the relative merits of every potential variable.

It is unavoidable on most projects for time to be wasted discussing set-up procedures, redefining quite obvious basic issues and the team bedding down into some logical cohesive group. Despite the Latham Report, *Constructing the Team* (1994) and the Egan Report, *Rethinking Construction* (1998), much of the construction industry is devolved, with individuals joining together for the first time as a team for each project. Making use of the collective experience is one of the most challenging issues for the industry. Procedures are required that actually work, not just pay lip service to the process.

Theory versus practice

It is often very difficult to put a message over to clients when they have been 'sold' a particular point of view at the start of a project. Over the years I have had many conversations with clients determined to use a particular material, technique or system. Sometimes they deliver what they offer, but all too often they fail to do so. New products, or materials from unknown sources, can be notorious in this respect. The roofing market is full of materials from all over the world, with those that are cheaper but that look the same as more expensive versions being the obvious front-runners. In such circumstances, the specifier is on the back foot from the beginning. Challenging an obviously brilliant material sourced from a South American company and shipped halfway around the world is very difficult if it is substantially cheaper than the mainstream European equivalent. Those who dare to question the track record, the specification, or the lack of tangible samples and examples are often seen as the villains of the piece, obstructing a perfectly sensible approach.

However, there are obviously risks involved – a cheaper material normally means a lesser specification, with a good chance that it will not last as long. But often the theory overtakes the practice and the cheap option is chosen. Almost as consistently, failure follows hard on the heels of the completed project. The net result is a substantial repair and remedial programme not long after the building is complete, ranging from complete reroofing to running repairs over many years. This is not limited just to the fabric and architectural elements – often it also applies to services equipment not delivering the performance so readily touted at design team meetings.

From direct experience, I would say that at least 40% of all mechanical and electrical plant does not deliver the performance discussed at design stages. This has an obvious implication for the whole life of a building – starting from a negative position, it will never get better, and over time will result in a major deviation from the planned, programmed path.

Conclusions are counter-intuitive

Consultants have not always been the best at providing the appropriate information for their clients. This often results in counter-intuitive answers. Frequently, the refuge is in the price rather than in the performance. When debate centres only on cost, there is no doubt that elements of the project will be lost, one of the first to go being whole life values and costs.

If all our experience tells us that there is no conceivable chance that maintenance will be carried out correctly, on time, and to the book, why does this principle continue? Most other industries have moved on. In the field of engineering, cars, boats and aviation have long been designing for the best and planning for the worst. In these areas, planned, programmed maintenance and replacement work is built into the culture. There is very little room for manoeuvre, therefore the designed outcome has a good chance of succeeding. Many clients are demanding better performance – though not, it seems, in the performance of buildings.

Before any sort of meaningful whole life conversation can begin, this situation needs to be changed. It is clear that pressure on resources and efficiency drives will affect this, but it is questionable whether the construction industry will ever mature to the extent needed.

It is no wonder, therefore, that clients make the decisions they do, bearing in mind the mixed messages they receive, and that their motives are far removed from the debate of price versus performance.

What you put in is what you get out

On the large project involving many buildings described above, we designers specified a flat roofing system to the client's brief, which required a 40-year life. Our initial consideration was to look at a metal finish: lead, zinc and copper can all achieve that level of performance. The location of the project was not especially complex, the exposure was moderate, and the building was relatively simple. However, any thought of using metal was wiped out in the first discussions on costs. Metal would be far too expensive based on purely capital cost alone. Such questions tend to focus on the unit rate for the building: if more is spent on the roof, there will be too little for the walls and windows. Over hundreds of projects, the client's aspirations and the cost plan never seem to match, resulting in the destruction of the design in order to achieve the cost plan.

Single-ply membrane was our next choice, but brings its own difficulties. The brand leaders do have the ability to perform for 40 years if installed correctly and not abused. They have a remarkable durability even if not maintained that well. But single ply is a crowded marketplace, and ensuring that one of the top performers is used is a further huge battle. There are over 100 companies worldwide producing single-ply membranes, and the spread of quality is enormous, so ensuring that you use a material that is fit for the job is a major undertaking. Single ply also has some negative points in the sustainability area.

The debate went on for far too long, and as in so many projects the real issues were in danger of being lost in the politics, bidding and counter-bidding of a large project. In the end, when time runs out, decisions have to be made – often at the expense of logic. So it was in this case, moving away from the best specification, the best value for money and the legitimate choice. The roof was actually built with three-layer felt – the traditional solution, with an analysis that the roof top sheet will be replaced at the half-life point. I have absolutely no confidence that this will work, but experience tells us that this will be long forgotten before we reach the supposed half-life in 20 or so years' time. There is a strong likelihood that the roof will fail, and then attention will focus on repairs. Repair work will involve disruption to the building users, and access to the roof with associated health and safety risks. In all probability, more work will be needed than if a planned and programmed approach had been adopted.

The alternative would have been a long life, less maintenance, the overall resources used being less, and the whole benefit for the client and the building being more positive in every sense. But this is the whole life dilemma – how to break this pattern of events and achieve a desirable route is a challenge the whole construction industry needs to consider. Keeping the *status quo* is not a sensible, ethical or tenable option.

Chapter 18

Redefining the future

If you don't know how, how would you ever know what would happen?

How hard can it be to set out rules and conditions that will govern the outcomes of the procurement and construction process? Is it really that difficult to ensure these are then used by building owners and occupiers for the majority of a building's life? Apparently, almost impossible at the moment. Experience tells us that this is a big ask. Rarely is there any engagement in this process, but I remain optimistic that this can be changed.

The object of this whole endeavour is to better inform procurement and construction actions in order that the final product will match the client's requirements. That may mean the building is tuned to the life needed and will result in minimal upkeep, or it may mean a short lifespan – but what is required should be delivered, rather than something randomly created to a different plan.

However, the way we go about this whole process inevitably will not lead to these outcomes. It is clear that the various systems, databases and predictive techniques available will all fail if analysed hard enough. This is because none of them has the ability to look at every issue in sufficient detail, and they do not offer enough control over the process to ensure everything that is planned is undertaken in the manner required.

Unlike Tom Cruise's character in the movie *Minority Report*, whose job is to catch criminals before they commit their crimes, it is not yet possible for us to have a clear view of all possible futures. If it were, then whole life costing would be easy. But at present we are locked in to the imperfect and frequently unpredictable world we have to deal with on a daily basis. It is therefore clear that any amount of analysis and application will not bring us closer to ensuring the outcome we want will be achieved.

'Factoring' seems to attract a lot of interest in this regard. Attempting to factor into an equation all the twists and turns to achieve a predicted life is backed with some enthusiasm. However, there is no accuracy or logic in this approach.

New ways of working

Ironically, 'new ways of working' is not a new phrase, but it is questionable whether they have ever been achieved in the construction industry, where generally the prevailing arrangements lurch ahead, occasionally benefiting from one small step and then another. The biggest motivators for change have always been economic or (very occasionally) political upheaval. During these periods, significant contraction or expansion of the industry will force a rethinking or restructuring of the established procedures. Outside these periods, it is very difficult to see how advances of a significant nature can be achieved when there is so much resistance and so many obstacles put in the way of any logical advancements.

Fundamental to this is the structure of the industry itself. Its compartmentalised and siloed structure gives rise to agenda-setting, fixed positions and the 'not invented here' principle. This always puts the brakes on any new way, no matter how laudable. Building momentum for change is crucial, but always a mighty uphill challenge.

The combined forces of the Latham Report, *Constructing the Team* (1994) and the Egan Report, *Rethinking Construction* (1998) have made some difference. Their highlighting of the issues and the need for change at least raised the debate and generated a reference point for change. However, many of their points have been lost or locked up in the machinations of the industry. Although some issues were taken out of the closet and openly debated, has a great deal changed?

So it is with whole life costing and value: many principles, many debates and a lot of paperwork, but equally many questions, and above all has any progress been made? Where are the real, tangible, provable benefits and rewards? Advocates say they are there, downstream, and will happen; dissenters respond that this is 'jam tomorrow'.

Will we ever change? Possibly. But we have to have a robust model to use and to trust each other, rather than a collection of 'wing and prayer' basics or a statistical solution accepting the *status quo*. Everyone feels their tried-and-tested methods are the way to go. But the great majority of projects are based around a cost-prediction model that is flawed. The detail and rigour are rarely there; instead projects tend towards a principle that detail, although really needed at the beginning, can be sorted out as the project progresses and finalised in time for the end of the project. This does not help whole life projections at all.

We need to have a much more complete understanding. It is very disappointing that the current estimate, tender and package principle delivers such poor results. Often detail on the cost is not agreed until far too late in the day. This has the net result that savings always have to be made, and as if this is not enough, profit must be wrung out of every stage. This is a mechanism to ensure change does not happen. It is also the major barrier to a consistent and appropriate specification, and will have an effect on the workmanship and performance achieved. This, then, is the single biggest issue when it comes to delivering whole life cost or value.

There are a number of routes that will take us forward, and any discussion of possible futures has to confront the issue of multiple futures. While the desired path may be prescribed, keeping to it and achieving it can prove very difficult. The number of possible paths, and the points at which a change of direction is possible, are numerous.

If we relate this to the principle of whole life costing, managing to keep the scheme of things on track and achieve the desired outcome becomes highly improbable. The whole life value process must attempt to filter the choices or possible paths and try to keep the desired objective in sight. This is by no means a simple or easy objective. The need to keep to a fairly narrow track is very important, and the possibility of success, by any measure, depends on this outcome. Sadly, the chances of this actually happening are remote.

Possibility/probability

Gamblers know all about the probable odds of one outcome versus another. The art of the possible is always balanced against the probable. A central question within any construction activity, asked at many design team meetings, is 'what are the chances of that happening?' But no real analysis is carried out to test the odds. Bringing into any discussion a debate on the odds of this or that happening is rare and always a challenge. Optimism often rules beyond reason – just the simple principle of staying on programme is a major challenge. When we consider the chances of one outcome instead of another, this is gambling: weighing up the odds and accepting the risk that the chosen route is going to be acceptable, even though we are probably not in control or even aware of all the variables, and have no real grip on the chances of reality matching our chosen path.

Alternative paths

We think we know the future, but more often than not this is an illusion. Because we plan and organise most of the time, we inhabit a world not fashioned around what we have determined, but resulting from a multitude of interactions. Because our view is limited, we cannot see either what might have happened, or how close our experience was to being completely different. It is difficult to prove a negative – what might have been, as opposed to what is.

The whole life conundrum is that it will not work because there is no faith that it will work. Many examples miss the point, and perhaps this is because we still do not understand the fundamentals at work within the whole life of a building.

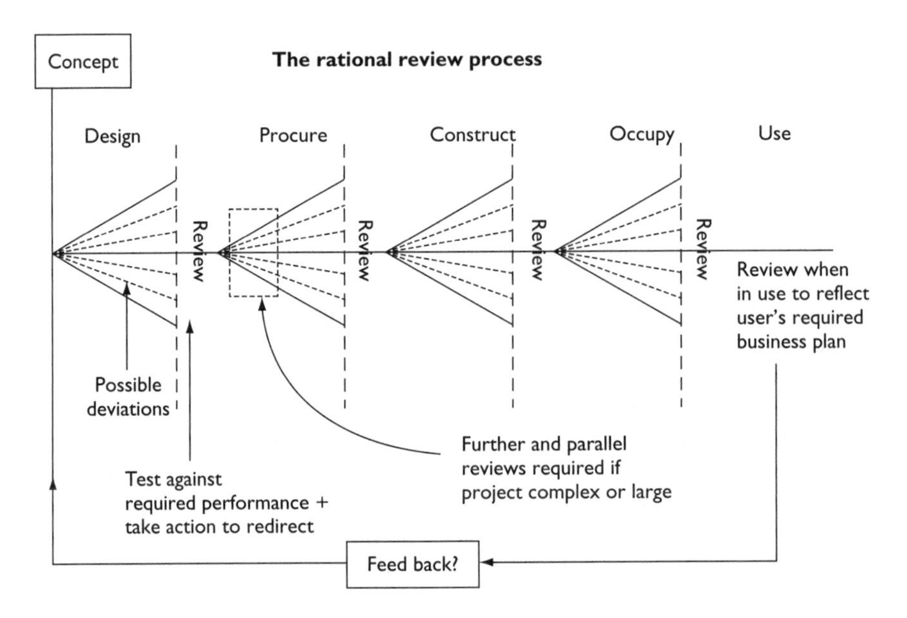

Figure 18.1 Keeping on the right path with gateways to correct deviations

We could compare this with man's early attempts to fly. Before man could reliably take to the air, many attempts failed. The approach was based around copying birds. Pioneers then started to look beyond simple analysis, and instead of copying birds as observed, they started to ask questions about how they were able to fly. Once the principles of flight began to be understood, reliable, repeatable performance began to be established. From this developed the refined performance of recent decades.

So it is with whole life values and costing. As with the fundamentals of aerodynamics, we need an approach that embraces and understands how the principles will work. Currently there is no evidence that this is anywhere near being achieved.

Chapter 19

Searching for the ideal process

Is there ever going to be an ideal process? Perhaps not – but we can do a great deal better than at present.

We need to look at how to match the issues of practical control with the benefits of whole life cost and analysis. As has been demonstrated, this is currently a challenge. The industry, the constituents, the protocols and the current methodology all combine to prevent this taking place. Perhaps it is not so much an ideal process that is needed so much as a direction leading to better methods that may improve delivery in time.

So where is the sensible line? Where is the logical pathway to drive this forward and ensure we can advance these issues without embarking on another fruitless and resource-consuming methodology that, in the long term, will not work or deliver any tangible benefit that is apparent above other, financial considerations.

A combination of periodic review and audit will bring some benefit in the direction we need. We know that a project can be procured with whole life value in mind. Possibly, it will be built close to these values, but it will not be maintained in the manner prescribed. There is currently no focus or desire to put any effort into this, almost certainly due to the lack of proven benefit.

It is possible to establish a set of criteria to provide a framework for the review process.

Design

First, the building's nominal life is agreed with the client. This is superficially easy, but in fact poses a real challenge. What does 'life' mean? What will be the client's 'on-the-ground' requirements? Not all parts of a building will or can have the same life. Is it just the main elements that are of concern; or should we take in as much as we can practically look at; or will we not see any real benefit unless the whole building is assessed? This is complex, but taking one step at a time will bring results. A formula can be agreed and built into the project requirements.

The building's expected use and performance should be agreed with the client. Allied to use and performance, there needs to be a definition of the range of these – a clear band of what the building will be called to do, outside which the whole life values will not apply. This is needed to bring certainty to the process and perhaps to inform checks on future operations.

We then have an agreed process and protocol for driving the design. This can ensure the materials, the details, and the way they are designed to be combined can deliver the whole life cost and value that are required. This should establish a baseline for the project.

Procurement

We need to lock these client principles in during procurement. The key areas need to be red-lined and identified as critical to the whole life value or cost outcome. Any change to these needs to come from the client after a full impact assessment.

This overarching structure and control needs to exist and to be checked at every point in the progress towards starting on site. Limits need to be imposed and checked.

Construction

The same is true during the construction process from the start on site. Clear controls are needed to ensure the prescribed standards are met and what was required is delivered.

Periodical review, analysis and feedback are crucial, and at every stage this is where direct cost can accrue due to the need to ensure accuracy. Checks need to be made against the brief, the design, physical progress and emerging details. Closeness of fit to these is essential.

To make this work, a control regime in proportion to the project is needed. For the average project, I suggest monthly reviews: during construction, six months after completion and handover, followed by 1-, 3- and 5-year reviews would suit most projects. This will change if the building is not occupied directly on completion, which may often be the case in the commercial world.

The principle of the review is to check that the building is still on course and, if not, to make adjustments to the day-to-day running to bring the project back to the expected position. There may be budget issues, and it will be up to the client or the building manager to keep the project on course.

This is all about outcomes, and it remains to be seen if any level of control can deliver the needed result. But a tight, regular arrangement has at least some hope of success.

Use

Day-to-day use, while having an incremental effect, is extremely important to this process. With better IT systems it is now feasible to build in a data-logging system to ensure life-value issues are identified and the building's operators are made aware of these from the start.

Building up the picture, so that actions required have been in the plan from the design stage, is essential. One underlying theme is that not enough accurate data are available. Arguably the insurance industry data is the best set we have, although this is based on failure, and comparing these data to the progress of elements through their life is difficult and certainly carries no guarantee of accuracy.

Maintenance needs to be much higher on the agenda than currently: at present it hardly features at all unless something important breaks. Then usually it will be a case of too little, too late.

Maintenance is continuous and ongoing. If it is properly planned and organised, there should be very little unexpected failure and all components should achieve their optimum life. Of course, repair will still be required: nothing is perfect and some work will be needed. However, this too should be part of the plan from the start, not a spur-of-the-moment reaction to a crisis.

Demolition is also part of the problem. Often the components are scrapped rather than reused, as the cost of reuse is higher than that of replacement. This is true in a world where resources are still relatively cheap, but we may see the value of certain commodities rise where this state of affairs is reversed.

Decoupling the money

Perhaps the greatest disincentive to the progress of whole life cost and value is the link to monetary values for any analysis. In any system going forward, it is necessary to break this link. Financial aspects are, of course, important, but they are just one part of the equation, and the other key constituents should also be given proportionate importance and respect.

So the next part of this process is to give every element, every action and every resource a cost – but adding to the plan, not driving it.

Building MOTs

There is every prospect in the UK that 'MOTs' or similar will be part of regulatory requirements in the near future. This will have a very positive effect in encouraging whole life value. The Energy Performance Directive has been in place for several years. Increasing pressure to ensure its implementation will also help. Further development of this into regulations will encourage the upkeep and planned maintenance of the building stock.

The European Building Performance Directive (EBPD 2) is on its way. Clearer, more practical and easier to implement (so its designers tell us), it will require proven performance to be established across the built environment. Hopefully at last we will have a mechanism to ensure performance in use is treated seriously. This will ensure a focus on life values and costs. As a part of the plan, the requirements in legislation and the framework established by standards will aim to make the process as robust as possible with contemporary technology.

Furthermore, energy performance certificates with teeth are a very real prospect required by the current draft of the Directive. Although requirements are often diluted prior to becoming law, it is undeniable that pressure to improve carbon emission performance, rational use of resources and efficient financial management should add pressure to help this progress in the right direction.

Chapter 20

Whole life costing: is there an end game?

The future need for some rational analysis is obvious. If we are to close in on the issues of sustainability, health and safety, we need some forward-looking protocols that work. At the moment these are sadly lacking low-carbon results. We need to look carefully at the threads of new thinking that are showing the way and perhaps we will come closer to a process that gives some logical benefit and even may get us where we need to be.

We need to consider carefully the components and methodology for the model, and make some significant advances for the better. Several methodologies are currently being used, with varying degrees of eventual success. The pressure to make real progress will be present only when it is obvious – when it is too late.

We need to use the tools to hand. However, they come with a health warning – the well known phrase 'rubbish in, rubbish out' applies here. While we have methodologies to analyse the situation and feed in data, for this to work the plan has to be followed explicitly.

To guide us, we have the International and British Standard BS ISO 15686, 'Buildings and constructed assets – Service life planning: Part 5, Whole life cycle costing'. BS ISO 15686 is a useful tool but, due to its ISO status, much of its content is a compromise. It usefulness is therefore limited and it must be applied with care.

The Standard was published in 2008 and developed through BSI Technical Committee 500, and included 24 interest groups. It was designed as a tool to aid surveyors producing life-cycle costing plans. The full standard is in ten parts, with Part 5 detailing life-cycle issues.

The approach adopted by the committee was to look at two paths of data. Information from the testing and segregation tool and the service life data is used to estimate the life of the material or element under consideration. Factoring of life conditions is then added and combined with the other data to produce the estimated service life.

This is a perfectly sensible method of establishing theoretical service life. However, the assumptions made are all based on a series of data that, in turn, builds assumptions into the result. In addition, there is no allowance for maintenance (or more precisely, the lack of it). Work is now being considered regarding this, but it is staggering that this has not formed part of the Standard from the outset.

For the Standard to be of any real use, the errors in the assumptions need to be cancelled out. A clear picture of maintenance use and abuse needs to be considered. For these reasons, this cannot yet be considered as an appropriate or accurate tool.

If not this, what can we use?

It is clear that a more rational and logical approach is required if we are to gain any benefit from the efforts that are currently being applied to whole life issues. Moving forward in the right direction is a challenge.

Adopting a milestone or gateway approach with teams during construction and building owner–operators does have some merit above the current systems. The major problem is understanding and engagement. Those most directly connected with the building are not engaged with the whole life process – where is the pay off for them? The benefits and financial models do not favour this approach, and there are many disincentives to any sort of dialogue, let alone action. With such a complex and drawn-out timescale, this is hardly surprising. However, there are real benefits to be obtained, and these need to be made clear.

If we took the same approach to all parts of the built environment, it would be in a very poor state. Milestones, gateways, and keeping the faith that this system will deliver results offer one possible answer. There is no doubt that we can set up a project with clear objectives and timeframes that will see whole life costs and values realised. However, the nature of business and society is that these fall by the wayside.

Where we go from here is critical to the future of the built environment. There is currently much discussion, analysis and effort expended – but is it to any real effect? The theory is that this will produce better buildings, and better tangible results for clients, building operators and users. The risk is that it will not. While there are faint glimmers of hope, where this may lead is still anyone's guess.

Several members of the whole life cost-value camp fully believe they are doing a grand job and we are building for the future. Equally, others regard this as a lost cause and will never acknowledge that it is anything but swimming against a very strong tide. For the optimistic camp to be right, we need some major changes. We need to build what we have designed and maintain it correctly; to do otherwise is to settle back into the familiar *status quo*.

We can paint a picture of the future that does start to explain why the energy put into whole life costing may not be entirely wasted. But it is still a leap of faith, and it will require more than logic and trust to see tangible results. We live in an instant society, in a moment-by-moment reality, and we are used to seeing cause and effect – if we cannot, doubt starts to creep in and attention moves elsewhere. The problem with whole life value and cost is that it is very long term. As a society, we are not good at this sort of objective, especially as it occurs with issues that may not directly affect anyone in the short term.

Statistically, we are not in a good place as the odds against the work being beneficial at the end of a project's life are very low. So many factors come into play, it seems almost inevitable that this approach will not succeed in any meaningful way, so why bother? We have a huge need for intelligent use of resources and buildings, and the effort put into them to provide long-term and viable returns. What we are really seeking is a better return for the effort involved in the first place.

I believe that the pressures from sustainability and carbon reduction will have an effect here. From a regulatory position, we are beginning to see waste and the effects of excessive waste slowly being outlawed and becoming increasingly more difficult to ignore. From a fiscal point of view, it is clear that costs must increase to pay for this.

The question is, will these two pressures together serve to push us more towards a sensible consideration of life values?

The availability of new technology may also be a help here as we see the steady take-up of building information modelling (BIM). This is the use of computer models to replicate the whole project in every detail prior to the start of construction. Its advocates see this as a way to identify any problems, order exactly what is needed, cut out waste, and program the project to be constructed exactly to time. There are also many who feel this is a little idealised and, while it is a help, it is not a universal panacea. Many problems remain in moving from the conventional to the BIM process, in particular persuading the client to pay for a computer model, and addressing issues of liability.

However, it does give us a view of a world that can identify the detail of what is to be built and how. This is a very great advance over what we have at present – which in many respects is as accurate as 'point at the target, close your eyes, pull the trigger and hope for the best'. I touch on this and its potential later.

The insurance world

We cannot look at this without considering the impact of the insurance world. Insurers are interested in risk – they manage risk, take on known risk, and review it. They consider the asset in many ways, but with respect to the outcome of whole life cost and value there has been little engagement, and rightly so.

It would seem likely that insurance would be a prime driver in ensuring the risk involved with a building into the future is logical. However, normally risks can be managed or offset, causing few or no commercial problems. If we are looking for drivers for the building to be managed throughout its life, then perhaps the insurance industry is an ally. There surely can be an advantage in the set of individual policies used to cover an issue such as roofing being wrapped up in one insured risk for the whole project from start to finish. There must be some benefit to the insurance company in ensuring that a project's life is being managed and the objects at handover are taken into account, reducing costs and pollution.

A further outcome might be that there would then be interest in setting up the project in a realistic way in the first place – that is, deriving a specification and ensuring it is adhered to. This invariably would result in the build cost increasing, but would enhance the value of the building as it would last longer and perform better. Our problem at the moment is that we rarely measure either in overall terms or in detail.

Issues such as fashion and commercial trends tend to override whole life aspirations, so the projected return on better quality build and maintenance may be destroyed by factors outside the project.

PFI clients are in a bit of fix as they bought in to a process where whole life cost and values were built in. However the project risk has shaped up, there will be someone who has taken on the liability and will need to address the issues. We are already beginning to see that some of the earliest PFI-generated buildings are a nightmare to maintain and operate. Early evidence suggests that they are difficult to run, likely to underperform, and suffer maintenance issues earlier than designed. I would say this is to be expected, and there is no consolation in suggesting that this is the tip of a very large iceberg. It is disappointing that this approach has persisted for so long without proper understanding of the long-term errors.

We also have the emergence of whole life carbon accounting. This may well bring new life to the whole life approach. The need to ensure that we control and account for carbon, especially in the built environment, is being given significant weight. The carbon input and use of any part of a building is obviously directly related to its life and function. Materials and systems that perform well, and for longer, make better use of the carbon investment as well as the fiscal investment.

The politics of whole life costing is not a wholly negative force, but it is largely to blame for slow progress. A driving force in the whole life and whole life cost arena is the political weight given to this subject. This is politics at both ends of the scale. Big politics sees this as firmly part of the agenda to ensure projects are appropriately sourced and constructed, and achieve social payback.

Arguably, politics with a 'small p' is also influenced by government authority, but there is also a sense of doing the right thing – of making sure that all bases are covered. At a detailed level, it seems that in public sector procurement, whole life costing is a significant subject despite the obvious pitfalls associated with it. There is consistently a significant push to ensure that all construction projects are given this analysis and that documentation is produced to ensure the whole life values have been produced.

But to what effect, and to what end? Experience shows that in real terms this issue hardly makes it onto any radar at all. Advocates tell us that it requires a review of all the materials sourcing and assembly intended for the new building, and considerable effort goes into the production of this information. This does have one tangible benefit, that during this analysis, the characteristics of the specification are at least made clear. So often in today's construction industry, we do not know (do we care?) what actually is being built. The process may be so diverse and fragmented that only broadly do we know what constitutes the actual build. Often where supply chains are three, four, five or more levels deep, despite a heavy paper trail, actual facts are few and far between.

However, the fact is that in order to progress, information on whole life costs and values is required to be produced and filed. I have no evidence at all that once produced, any of this information is used to any real advantage. On the contrary, there is strong evidence to suggest that after submission to the client, it is never looked at again.

In the same way, many other issues cloud the process and contribute very little. There has been a clear development that the inclusion of a whole life cost is required and will be good for the project. I suspect, however, that the establishment of a decision-making analysis is seen as an end in itself – it could be called the 'justification document', there to identify that all the decisions along the way were undertaken with no bias and in absolute accordance with the requirements. This is clearly a back-covering exercise designed to protect decisions, and possibly people, from accusations that mistakes were made, instead of ensuring that the project proceeds with clarity and the right actions are taken. No project is faultless, of course – there will always be mistakes. Many are working hard on better methods to reduce errors, and there is evidence to suggest this is working.

Over the past few decades, we have seen more complex, more daring and challenging projects, with varying degrees of success. In the main they have been undertaken largely in the private sector, based on good practice alone. The need may be felt to ensure political correctness, or transparency, although some would describe it as having exactly the reverse effect, often contributing to the very issues it is trying to avoid. In any event, we end up with considerable waste added to the project, for no benefit, and driven by the political expedient to be seen to do the right thing.

I believe that with new methods of looking at the problems, better data will result in a change of heart, and the politics of this will start to resemble a sensible proposition. Continuing with the *status quo* is to condemn a lot of construction to continue with pointless, never-to-be-seen documentation.

Why can we not have a better way?

To answer some of these questions, we have to disconnect whole life issues from the project, move in to identify what character the project will have, and then structure its life going forward. Having just a vague idea of the possible outcomes and linking this to statistical principles is not going to work. The paper-pushing bureaucratic world will always want to establish a back-up. This is correct procedure, but this does not help the project.

Instead, the project should be profiled, with an analysis of the quality required; the use required; the users anticipated; and the profile it will have for the foreseeable future. This should set the likely outcome path.

A number of life judgements can then be made – as designed; not as designed; and somewhere in the middle. A more developed and sophisticated version could allow for infinite degrees of deviation between these extremes.

We currently have a flawed but favoured system of whole life costing. We need to make this sorry state of affairs evident to the industry, the policy-makers and, most importantly, those who pay the bills. We need to explain in clear terms that there is no point in continuing to support this process and expect beneficial results that simply will not appear. There is not a shred of evidence that this ever will result in any real and tangible benefits; in many cases, the effort employed is disproportionate to the potential benefits. This is a shame as the original aspiration is both logical and intended to be beneficial.

What can replace this, if we move away from the whole life cost saga? A rational and deliverable system is needed.

Why do we need any system?

Leaving aside the political pressures, some degree of analysis is advantageous for many reasons. We need building users to understand that the performance of buildings in use is complex, unmonitored, and most of the time left to the lowest common denominator. If we are serious about this, we need to tutor those involved in building maintenance and operation to be efficient and to use the asset as designed, despite pressure to do otherwise. We need to cherish and hold on to the resources employed in the construction and use of every building. At present there is only limited pressure, including the need to curtail energy use, but the wider long-term picture is that there should be limits, and there must be care in applying those limits.

All projects must be better controlled, and that requirement does not stop at the point of initial occupation. Use of the Building Research Establishment's

Environmental Assessment Method (BREEAM), the US Green Building Council's Leadership in Energy and Environmental Design (LEED), and others can only help to set the project off on the right road. However, in their present form these are very much indicators rather than certain proof of performance in use.

Ensuring that the build employs the materials specified and is constructed to the required quality poses a major challenge. BIM may help in this, but it will need reorganisation of business models in order to work efficiently and to the benefit of whole life issues.

Feedback, testing, verification, sign-off and certification can establish a benchmark standard needed to track anticipated against actual performance. A programme of maintenance input versus performance needs to be employed, with a variety of levels (perhaps gold, silver, bronze). These would enable predictions to be applied relative to actual performance. Lastly, we need to consider the use of the building – is its use long term? Are changes occurring?

All of these tangible issues need to be factored together to establish an understanding of the likely life outcomes. Detailed consideration of a life path for a building can give some confidence that life predictions will match reality. Regular checks are suggested at maximum of 5-year intervals. The longer the life, the less predictable the outcome, and these checks are essential.

This is a long way from the current clear figures, but it is achievable and, most importantly, predictable.

Some questions for the consultant's team:

- What are the issues for clients?
- Where are the solutions from designers?
- Where are the solutions from industry?

Some questions for the client:

- How is this relevant?
- Are there more important issues to think about?
- Is this part of my business plan?
- What methodology do I use?
- Do I have a realistic mainstream budget and an occupation and maintenance budget?
- Does this make a difference to my business?

Key points

Steps that prevent success

To attempt a successful whole life evaluation and costing requires a fight against some strongly embedded industry issues, which is why we have so few practical models to explore.

Industry inertia

The UK construction industry inertia has long had a stranglehold on change. Any new system of principles takes a great deal of energy to ensure it is adopted.

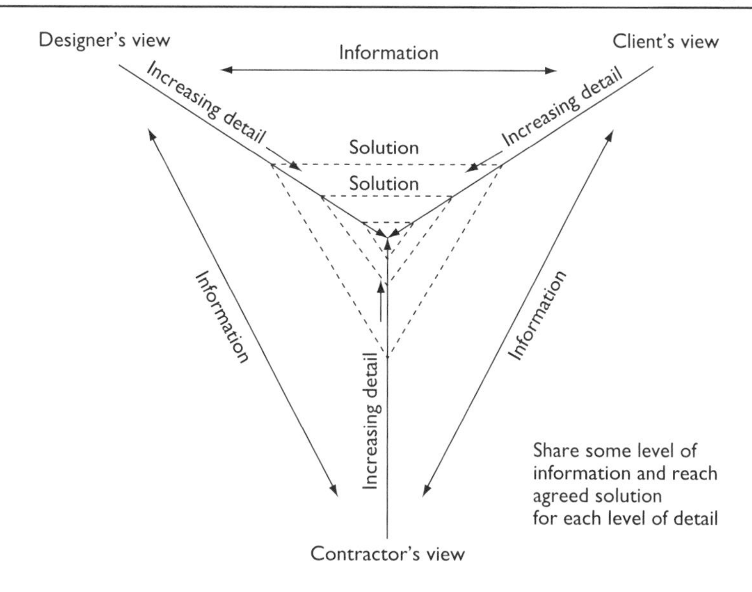

Figure 20.1 What's needed?

Lack of understanding

The 'I know best' syndrome is widespread, backed by the logic that 'if it worked in the past, why change it?'

Benefits are diffuse and over a long timeframe

To prove that the benefits from a whole life values system are, and have been, a positive asset to a project, it needs attention way beyond that paid by any current project team or client. The focus is normally on the here-and-now, driven by our short-term need for profit and tangible success.

Lack of credible data

Despite some good efforts and some notable databases that are available, it is necessary to ask, are these data credible and reliable? Hopefully with greater emphasis on proven performance and post-occupancy evaluation, the need for care and accurate understanding of buildings in use will be met.

Lack of proven evidence

While working as an architect, and in researching this book, I have not identified any clear evidence of whole life cost or value being put to work.

That is not to say we should not try. We should – if only to establish that the direction we have taken is correct. 'Live long and prosper' will take on a new meaning for future generations, we hope.

Chapter 21

The underlying principle: entropy

Ageing is defined by physicists as entropy. The second law of thermodynamics, in simple terms, dictates that all things move from order to lesser order. This is moving from low entropy to high entropy. In principle, without any external influence, any system, material or physical relationship will break down and, if given long enough, will revert back to the atomic building blocks that created it in the early days of the universe. On an astrophysics level, there is considerable debate as to how far this process could go and the potential effect on the universe as we know it. That is many billions of years into the future, a long way removed from the practical considerations of life as we know it. However, what we experience and attempt to control is governed by this principle.

On a more mundane level, the pursuit of whole life values and processes has to accept that entropy, like other fundamental laws of physics, cannot be stopped. Instead, we need to work with it to ensure the principles and procedures we put in place are harmonious with natural laws. This will result in a more efficient and satisfactory outcome in all respects.

This is a long way from considerations of whole life value, but it is important to understand that we are attempting to control and, in some cases, arrest the effects of this principle. We cannot arrest entropy any more than we can stop time.

It is always best to start with the basics. With that understanding, we can then derive means to predict, and perhaps control, the change in any system. The manner in which entropy changes gives rise to the concept of the 'arrow of time': a clear indication that time is a universal effect and has a similar effect on all things as far as we currently understand it (although there is still a long way to go in our comprehension of these fundamental issues).

Practical construction, use and maintenance processes are forced to accept that very little can be done to arrest the underlying ageing process. This overriding principle means that, when considering carbon, we are always going to be faced with materials breaking down, and with complex molecular bonds separating and reforming with simpler ones, affecting many materials used across the built environment. This is seen in the ageing process – the breakdown of components, the fading of paints, the discoloration of plastics, etc.

The carbon problem emphasises the need for us to understand these processes clearly. The central question concerning ageing is our ability to predict these processes and to make due allowance for them.

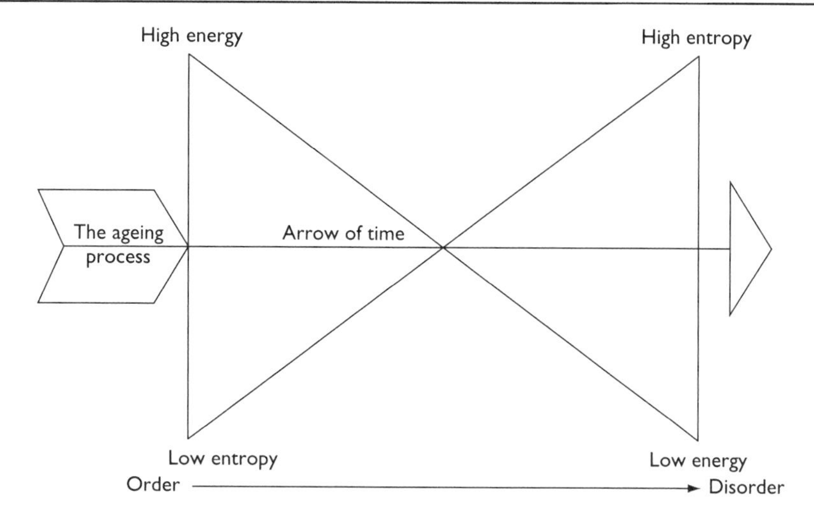

Figure 21.1 Bow-tie diagram

As a carbon life form, we are part of this process, and our own biology is part of the inescapable entropy issue. Continuous oxidation and transfer of energy is required for life. It is also the trigger for ageing.

The best we can do is to ensure the move towards disorder is controlled and appropriate. We are then making the most of the resources available to us, and ensuring that the whole life value is as high as it possibly can be.

Chapter 22

Carbon whole life cost

Whole life cost and value have languished in the slow lane for a long time. The first thoughts and ideas in this direction date back to the early post-war years. Throughout recent decades, a family of industry guides and processes have been produced, most stemming from government projects, broadly with little impact. There has been more interest in the past decade, but while there has recently been more optimism and constructive thought, there is no real momentum in the mainstream to adopt whole life values or whole life cost as a key project driver. There is a 'new kid on the block' – whole life carbon.

The link between whole life issues and sustainability has been discussed in previous chapters. A fundamental part of sustainability is the effect of carbon: intrinsically the two are part of the same question. Can we change the way we conduct our affairs to reduce carbon and conserve resources before it is too late? These questions are real and serious. Our intensive use of energy and, with it, the consumption of fuels releasing carbon and other pollutants is going to have a real effect on the world's future. This is serious stuff, and as the debate is played out, the need for us to gain a tangible grip on carbon emissions is certainly near the top of the agenda, if not the most important issue to be resolved.

This is seen as one of the key drivers in climate change and possibly responsible for global warming (although there is still much controversy over the exact roots of global warming). However, it is clear that increased carbon and the other gases highlighted are implicated in the changing atmospheric conditions we see around us. These trends have stimulated interest in carbon in all its forms. Arresting the increase and then reducing the levels of carbon in the atmosphere is seen as one of the highest priorities facing the world.

The relevance here is that carbon is involved in the majority of human activities. In the modern world, there is very little we do without using energy, and for the most part that means using part of the carbon cycle. It is rare that a process or activity does not result in release, in some measure, of carbon into the atmosphere.

Carbon and the effects of emission are therefore the subject of debate, research and not a small amount of effort to establish how we can deal with this problem on a global, national and personal level. Each one of these areas carries immeasurable challenges, and potentially will involve changes to the current lifestyles of millions of people. Politically, therefore, it is a real hot potato.

There is an urgent need to identify how best to minimise the effects of carbon across nearly all aspects of daily life. Change for the sake of it will not do – we need to make the right choices and the right changes. Just using energy in a different way will not always bring about a beneficial change. Often, when all the facts are taken into account, no net improvement is found, and suggested changes may make the situation worse. So-called low-energy devices, seen holistically, such as solar panels and wind turbines, often generate more carbon in their manufacture than they save when put into service. This only brings a short-term or superficial gain; in the round, it gets us nowhere. The overall reduction of carbon is very difficult to achieve.

Carbon in the atmosphere is created by power generation, transport, manufacturing and construction. The construction industry is responsible for a considerable proportion of carbon and may be seen as an easy way to reduce levels in the atmosphere.

One person's carbon emissions are another's embedded carbon – the interrelationship of who uses and benefits from what is a problem of global proportions, and it is taking the best minds to consider what can and should be done.

The carbon cycle and its effects

There is huge potential for change that will have major economic and political consequences, and there are many who oppose the idea that we can have a tangible effect on carbon use and climate change. Construction and building projects are at the forefront of energy and resource use. It is therefore not surprising that a focus on the amount of carbon generated in the production, use and maintenance of buildings attracts so much attention. Every stage in the procurement, construction and use of buildings requires energy.

- Production of the materials for use in a building, fashioning of natural materials, transport to the site, construction and infrastructure all consume considerable levels of energy and all use a carbon-based system.
- Once a building is in occupation, the heating, cooling and general operations continue to increase its carbon load.
- When a building reaches the end of its useful life, some energy will be required to change or replace it.

Understanding the whole cycle of the carbon used is fundamental for the future of the built environment. Attempting to reduce the use of carbon in this context is a considerable undertaking. We need to consider carefully how we account for this carbon in the future. When does carbon start its effective life in the cycle, and how can we minimise its impact?

Nothing is truly new – minerals and raw materials are collected from the Earth's resources and will eventually go back to the Earth at the end of their life. Except in some very rare examples, the Earth is a closed system, and many are now pushing hard to ensure that we accept and work with this fact, rather than behaving as if we have infinite resources at our disposal. Continuing to use energy primed from carbon will simply place too much stress on the biosphere, resulting in increasingly extreme weather conditions. Estimates put the over-use of resources on a world scale as three and a half times those available.

The carbon cycle – and how we use it, account for it, and adapt away from its use – is possibly the single most important technical challenge that the world faces. It is this driver that is forcing the analysis of the carbon whole life cycle to the top of the built environment agenda.

There is no doubt that accounting for carbon and the way in which the carbon cycle is compared between projects will be highly significant in all industries as the twenty-first century progresses. Our level of analysis for the moment is fairly simplistic. All the tools we have to hand are relatively crude. Above all, we do not yet have a common measurement system or universally agreed methods of analysis. However, various groups, mainly through further European and UK standards groups, are working towards these needs – these much needed tools are now emerging and will be published over the next few years.

Much of the analysis is shrouded in debate over how embedded and non-embedded carbon can be measured and accounted for. In the case of any particular material, where does the cycle begin and where does it end? The whole issue of cause and effect is still subject to review and analysis.

Even accurate measurement of the basics is hard to find. However, there is a lot more momentum over this issue than nearly any other, and it is likely that solutions will be found, even if this is an emerging answer and evolution to the ideal model takes some time. Such are the pressures that it is a near certainty that, before too long, a fully costed carbon cycle, or at least a common model, will be adopted. This will surely lead to a renaissance for whole life costing.

How we measure, analyse and compare carbon use is a very complex and real question. Measurement, and indeed the whole analysis of the carbon cycle, its generation and reduction, is still very much in its infancy. Some analysis indicates that we have made little progress. Across the developed world, it is a challenge that has promoted the development of systems and processes to do the job. But none has achieved universal acceptance or acknowledgement as yet, and few have made any impact.

Measurement is very difficult for two fundamental reasons.

- Where do we start, and to what level do we take account of the effects of the carbon? It is very easy to double-count, or to fail to include some that should be included.
- How do we measure embedded emissions and conversion? There is a level of sophistication required that we have yet to tackle in order to produce a methodology that could be used on a day-to-day basis.

Additionally, how does this relate to the real world? We have countless examples and theories, but the correlation with real-world values is not clear. Is there a hint here of the fate of whole life costing also happening to carbon analysis? That is, much enthusiasm to start with, quickly fading to nothing more than a paper exercise.

Whole life starting with 'creation'

The following simplistic analysis of the carbon cycle helps to explain the problem.

- Energy is released from fossil fuels, oil, coal, gas and natural sources.
- Release of these hydrocarbon molecules into the atmosphere occurs.
- This results in the creation of a different atmospheric balance.

The level of these gases has risen manifold in the past 100 years, and many now say we have very little time before the tipping point is reached where the consequential change will be unstoppable. The increase in global heating, caused by the greenhouse effect due to more carbon in the atmosphere, may reach a point where no amount of carbon limitation will prevent further heating. This is the so-called 'doomsday prognosis'.

Changes in the energy flow in the atmosphere are being linked to more aggressive and disruptive weather around the world. The changes driving climate change can now be seen, and predictions identify some serious challenges to come. Current thinking is that carbon increase must be arrested and reversed before the tipping point is reached. Disruption to all life is forecast unless this is achieved, making this subject one of the most universal and crucial issues since the threat of nuclear destruction during the Cold War.

Looking at a finer level of detail – what are the relevant issues in this process? Energy is released into the atmosphere at a continuous and frantic level by transport, industry, heating and cooling in housing and occupied buildings. These are the primary sources. There is also evidence of significant input from natural sources such as volcanic eruptions. Hence the drive to reduce or use other sources of energy generation in addition to hydrocarbons.

This new interest in the fundamentals of carbon use has also reinvigorated interest in whole life issues. It will be interesting to see if this new driver will revitalise the whole life cost debate and add the realism that has always been needed across the built environment.

The new currency is carbon

There is a cost that has been levied against carbon. At present this is woefully low, however, because carbon is the life-blood of the industrialised world. Too much change, too radically, will rapidly bring civilisation to its knees. We have to move gently but positively into a new world of reducing carbon use without dramatically destroying the way of life for millions. This can be done only through correct accounting and costing of carbon in all its forms. We are at the start of a new journey.

The construction industry and the built environment have a major part to play in this. Not only is carbon important in building use, but it is also a substantial constituent in the make-up of the built environment. Many of the high-mass materials – concrete, steel and brick, for instance – require substantial quantities of energy for their creation. Concrete, for example, contains cement, one of the most energy-hungry materials in its processing. Bricks need energy-intensive firing for many hours, and that process is responsible for proportionately large quantities of carbon emissions. All this energy comes from carbon sources.

The energy use is regarded as embedded into these materials because once they have been created, that price has been paid from the Earth's resources. How we use these materials, and in what way, is thus of considerable importance. Once they have been created in this way, we should follow the principle of preserving the materials we have, without involving more carbon in the process. We have, in effect, invested energy in their creation that is borrowed from the Earth's available resources. Without careful management, many of the Earth's natural materials will run out – and many already have, or are being depleted to a serious level.

Construction is responsible for a considerable part of this problem. Carbon, seen as energy, is wrapped up in all of the everyday process of building creation and use. Every process – curing, firing, refining, forming and assembling – uses energy. The invisible energy load of every building needs to be recognised and should form part of the overall understanding of that building's importance. How much energy and carbon has been involved in the construction process? How much will go on to be used during its occupation over its lifetime? Currently, these are very difficult questions to answer.

Quite apart from construction activities, transport and infrastructure also play a part. A significant part of the process is the need to move materials and components to the site from wherever they are made – nowadays this can be from just about anywhere around the world. This can add a considerable carbon load to any construction project, but as it is currently not accounted for in any tangible sense, it largely goes unchallenged.

On completion, a building starts its life as intended, immediately consuming energy and therefore contributing to the carbon emissions generated every day. During the building's life, these are due to the designed performance and the actions of the occupants. Research has shown that, despite an advanced specification, actual performance can vary considerably due to the behaviour of the occupants. This appears to be true across all building types. Wide variations between designed and actual performance need to be analysed and accounted for. Programs such as CarbonBuzz in the UK (www.carbonbuzz.org) seek to gather data on buildings in use in order to gain this understanding based on real data.

This lack of performance as designed is of considerable concern. While the carbon load due to procurement and construction is complex, at least there is the prospect of arriving at a formula. In the area of buildings in use, we are still at the very early stages of establishing the drivers and outcomes.

There is an unfortunate parallel here with the non-carbon whole life costing world – while much has been published, discussed and considered, it has little impact on buildings in use. The common issue is the lack of real drivers to get these issues placed high on the agenda. We cannot afford for this state of affairs to continue. Many governments have now published their carbon-reduction plans, and the UK government has been very proactive in this area, setting out not only a programme and a plan, but also details on how it expects carbon reduction in every area to be achieved. Plans have been published to establish how the built environment will be affected. Above all, the structure needs to account for carbon at every stage of the process. While there is much to be finalised, a clear plan, framework and projected outcome is now at least on paper.

But all this good planning and organising will not count a jot unless the hearts and minds of building users can be influenced to the point where they understand and contribute to the efforts required. When considering the carbon generated in use, there are wide behavioural variations making predictions really difficult. This issue will require a great deal of focus to unfold and provide a real-world answer. However, the need driving this issue is now becoming accepted, which will help the overall cause of whole life values.

I am concerned that, while there is considerable pressure to reduce carbon, and huge effort is being applied to that end, the issue will still flounder in the same way as whole life costing – with huge quantities of paperwork, hot air and theory, very little real practical application, and negligible results.

What is needed is to get a grip on all elements of a building's life. This is as important for the originators as for the occupants and, indeed, for governments. If we are to have any success, then all parties will need to show that they can contribute, and

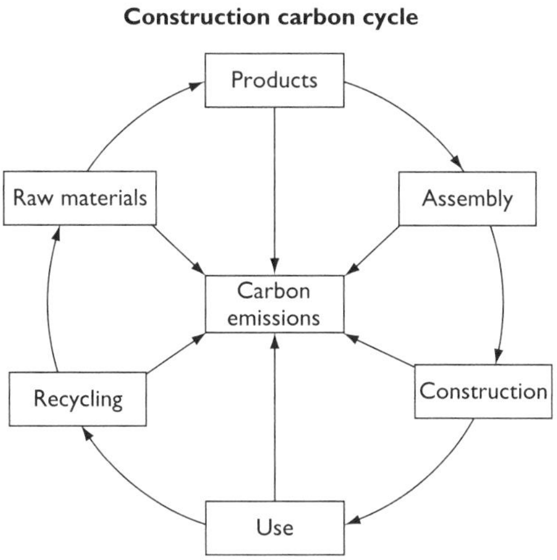

Construction carbon cycle

Products
Raw materials
Assembly
Carbon emissions
Recycling
Construction
Use

Figure 22.1 Carbon parallels whole life issues

show that they mean it. We are talkin of nothing short of a revolution in the way society behaves – but equally, only evolution will provide the answers and attitudes needed. Such is the magnitude of the changes needed that care must be taken to avoid creating problems of equal measure to those we are attempting to solve through over-zealous or draconian actions resulting in unforeseen and unwanted consequences.

Applying a practical and common-sense approach leads me to believe that we can set out at least some of the answers in a logical, and down-to-earth manner.

A formula for identifying carbon in use can be generated by starting with the basics using a reasoned formula involving:

- location
- building type
- types of activity within the building
- nature and intensity of use – initially defined as high, medium or low
- a factor element for detail (although I am generally against factoring as it introduces inaccuracy and can be just a 'fudge').

These elements will need comparative data to be usable, and work currently being undertaken in the standards world is attempting to define these issues. Using researched data for these issues will make the establishment of a starting baseline possible. But that will not be possible until we have a common expression of carbon cost. While carbon trading began some time ago, there is a lack of reality in the current levels of value. Whole life carbon is even less well defined or costed, and needs to come of age very quickly in order to make sense of all the effort being applied.

Someone's carbon emissions are someone else's embedded carbon – where is the starting point, and how can double counting be avoided? These are challenging questions, and only some of the answers are currently emerging, but there is no doubt that they bring new life to this subject area.

Chapter 23

Recommendations

Having considered the problems, I would now like to look at some possible solutions – practical recommendations for the real world. How can we set up a project to identify the carbon cost, and possibly to make a better attempt at addressing general whole life issues?

What elements are needed? This may sound obvious, but in surprisingly few instances is time taken to work out to a reasonable level of detail the main elements and drivers of a project from the beginning. As a minimum, the following are needed.

A project plan

To get this right, you need a road map that works and that everyone will work to. This is made all the more difficult because it has to work for years and years, not just during construction. Many will claim to have a project plan, although often it may be very thin in the area of whole life issues.

A project plan is need that is measurable, testable and usable as the framework from which to develop everything else – it is the foundation and the route map from which everything else derives. It must be clear, measurable, practical, and must anticipate the entire life of the building. This is the key – so much of what we plan and do is based on short-termism, this has to be the first philosophical change.

The plan needs to include the whole picture: design, procurement, use, upkeep and future versatility. It must satisfy not just the initial brief and client, but also future clients and owners. We will now examine these in more detail.

Client brief

As with many other aspects of construction, the client sets the scene. Without the client's buy-in and drive, the enterprise cannot happen. A project may have several clients over its life, all of whom need to have the same commitment and, preferably, direction.

A good brief sets the scene, giving enough direction without being closed to innovation, while making clear that the project's success will be determined by achieving its goals. In future, client briefs must spell out the carbon aspects required; many already do, although they may not yet have been delivered or tested. The aim should be to give clarity to the future shape and form of the project for decades to come.

Project documents and data

These are the vehicles that set the delivery standards. They are created from the framework set by the project plan and populated by the requirements of the brief. Quality is essential here. We see many different formulae applied to the generation and control of project documents, now supplemented by intranets and the increased use of building information modelling (BIM). All promise greater consistency, efficiency, and ability to ensure whole life issues are established throughout.

Clarity is essential – in the past, establishing exactly what the project documents determine at any one time has often been difficult, resulting in some success, but not without errors and faults. A lack of clarity causes mistrust, and in many cases causes financial frictions that affect confidence, trust in others, and the ability to innovate.

But today, cutting carbon and cutting costs are driving construction everywhere, and perhaps we have an answer that will deliver both. The world of the virtual prototype, built in its entirety in the computer before any resources are used in the real world, promises much. Having all elements identified, not just their physical characteristics but also in carbon terms, provides an opportunity for real carbon accounting. The obvious spin-off is that establishment of life values cannot be far behind.

In such electronic projects, every component is identified and has real values. Applying this in use, and for maintenance purposes, provides day-to-day guidance for management that could never have been achieved previously, and hopefully will help to deliver real-time life values.

It is not too difficult to imagine increasingly joined-up working between the electronic data set for the building and its real-time operating system. As the project matures, real data replace theoretical design data, improving the ability to match design and performance, and making it possible to identify when performance is falling outside the design parameters (perhaps by increasing carbon emissions or reducing the life of a component by a currently unmeasurable amount).

We have seen the emergence of controls in cars that can take on the immensely rapid analysis needed to determine engine management traction or stability control. It is not impossible that the complexity and power of such systems could take on the analysis of a building and check all the components to provide sensible real-time information, allowing early corrective action and keeping the building within the required parameters.

References and data sources

Good information is fundamental, and in many areas we lack the data needed to make logical and sensible decisions. However, the world is catching up fast. In the next few years, data will be available that will have some hope of delivering the knowledge needed. But this is not to say we should relax – quite the reverse. Some success in this area should spur on more research to ensure we can take advantage of better, clearer data and ensure better performance in years to come.

Good and universal procedures, which everyone buys into, come from clear data. In every walk of life, we need solid data to deliver systems that everyone understands and can use in a practical way. The change of mindset we need in respect of carbon should be brought about through this process. As better, clearer data emerge, the development of simple low-carbon technologies and lifestyles will emerge. To make a change, the principles and processes must be clear to those affected. We therefore need to be robust and sure of the data, the methodology, and the practical means of implementation.

Solid reference sources are hard to come by. The data sets that do exist are limited by their scope or depth. But in time this will improve and we will have methodologies that do provide solid solutions.

Project set-up and structure

There is a need to get a grip before the project commences. All to often, a project heads off down the road without real analysis or direction, based on industry 'norms' and without careful consideration. This is where it all starts to go wrong. Without a clear establishment of the parameters, there is no defined starting position.

In order to correct this, a process of baselining the carbon content of the project is needed. This will need careful analysis and some interrogation of the detail. It also relies to a large extent on good, solid data being available. Base-lining, in these terms, is the process of looking through all the project requirements and establishing their carbon content or potential carbon load. This includes physical materials, electrical and mechanical systems, and the time and distance these need to travel to arrive at the site. This establishes the relative worth of all the constituents.

Having identified the baseline, the project can then progress with this as the reference point. Going forward to the detailed design prior to procurement, several further reviews will need to be undertaken to consider if the project is on track and, if not, address the issues pushing it off course. Reassessment may take place prior to tendering; prior to start on site; and at completion of construction. These data then need to

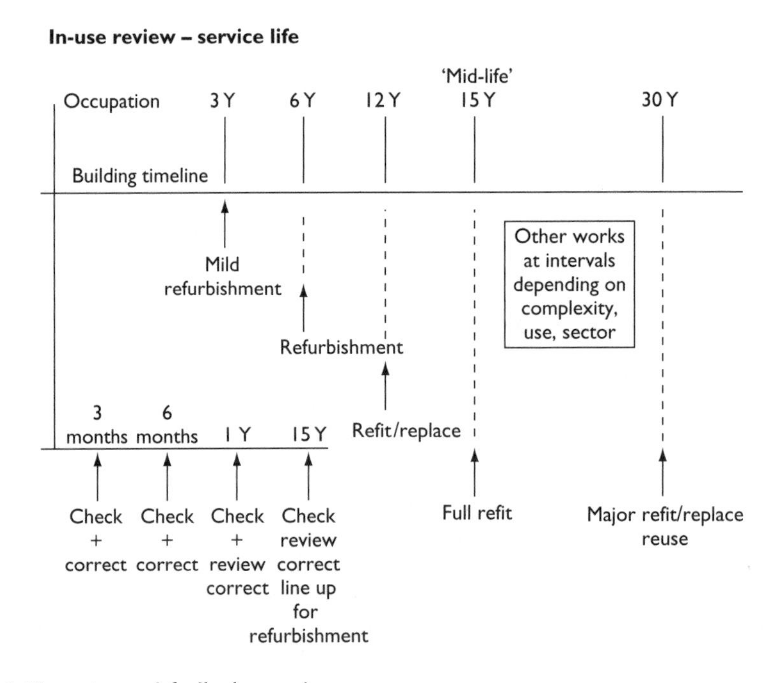

Figure 23.1 Key points and feedback to each stage

be reviewed against the original track, always with the aim of improving on the baseline situation.

An essential element is a review at the end of fitting out or commissioning, to ensure the building gets off to a good start and the performance is well understood. This should be followed by reviews during the building's early life, say after 3, 6 and 12 years, followed by a mid-life review after 15 years and an end-of-life review after 30 years. Most buildings will undergo substantial change at this point, if not before, and a revised baseline should then be considered. In the same way as the service record for a car, such reviews help focus attention on the building and its performance in the right way.

BIM

Building information modelling is not just a system of producing the whole building as a virtual prototype in a computer database, it is a different way of working creating collaborative framework for the project. This concept has been around for several decades. Used with a varying degree of success, it holds the hope of a dramatic change in the way we produce project documentation.

BIM enables the whole team to contribute to one information database, i.e. fully collaborative working. This develops in detail and shows all the relationships of materials, fixings and fittings. Every element can be seen in correct relationship with every other, making very clear what is to be built, and ensuring that any problems can be resolved early in the process. In its full application, the model is used by the procurement team to order and call off materials because the specification and physical properties are also available for every element.

This also makes possible the use of these data by the full supply chain, to the extent where factory production can be controlled by the same model. This is a fully joined-up process. On completion, the model can then be used to assist with maintenance and repair.

An interactive database giving access to all the information as the project progresses is what whole life analysis needs. All the information concerning the carbon content and values can also be included. Changes in the specification that may affect long-term values can be reviewed and considered. When the building is in use, the model can be used in verifying performance and, most importantly, checking if the carbon emissions performance is as designed.

This level of control over a whole project has previously been available only in theory, or in isolated pockets of innovation across the world.

The use of BIM is now being promoted by many governments, and its use in mainstream projects is being actively pursued. The ability to be able to consider the effects of changes still requires a process built in to the project plan. It still requires a team with the knowledge and experience needed to consider the options and advise the client. It still requires careful balancing of the options and sticking to the principles of the plan if real whole life cost analysis is to succeed, but it is a huge step forward in simplifying the process. Access to the data is crucial – ensuring that answers can be obtained easily in real-time puts the prospects of these considerations on the agenda for the first time.

Hopefully software providing easy analysis from a BIM model will be available relatively quickly, making whole life and whole-carbon assessment viable in one jump.

In addition to BIM, we now see more emphasis generally being placed on carbon-related issues. Nearly all environmental and energy analysis tools now have a section on

BIM (building information modelling)

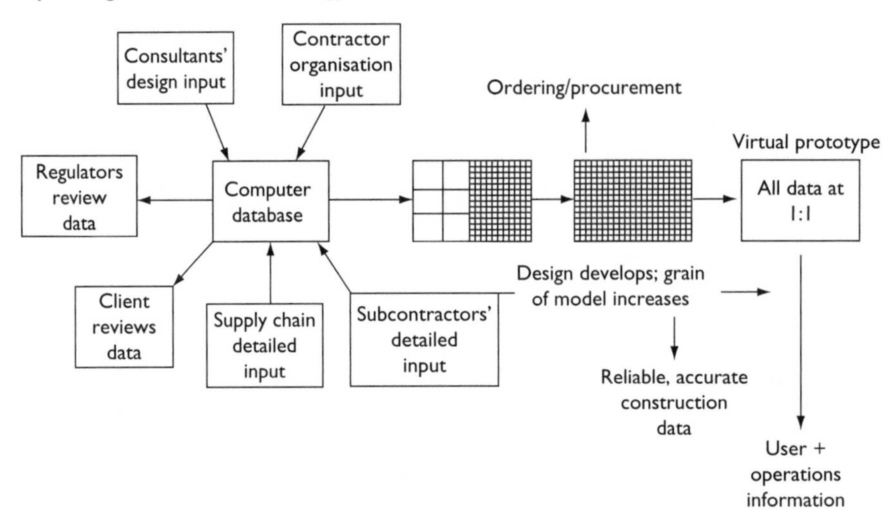

Figure 23.2 BIM working

carbon. This is encouraging, and the use of these with a BIM process starts to change the game in favour of a properly analysed carbon profile. This, in turn, will allow sensible carbon costing. This must be the immediate goal *en route* to the establishment of the carbon whole life for a project. While the principles are clear, there is a long way to go before all the information and systems are easily available in order for this to be a common approach. We can, however, look toward a joined-up approach that will deliver sensible carbon analysis, at least for the early years of a project.

For these issues to be worked through within projects, effectively they need to be embedded in the make-up of the industry. Consideration of whole life issues is sadly lacking in large sections of the current industry – while there is a lot of theory, there is very little practice.

I see little interest in large parts of the client arena. Consultants and contractors are, in the main, ambivalent, although they will grasp the subject if required to do so by the client. But above all, this whole area withers on the vine during occupation and use. Even the greatest exponents of the whole life proposal admit (albeit privately) that there is no take-up and little appetite.

This is why ensuring we have both embedded and free carbon under control is bringing new life and interest to the subject, although currently it is not part of most thinking about environmental issues within the industry. The drive to look at carbon at every level is forcing consideration of real carbon values to become a major part of the current agenda.

We are therefore on the verge of a sea change. For as long as I can remember, using fewer resources and less energy, reducing waste, and ensuring we maintain the environment for the benefit of future generations has been discussed endlessly – but has advanced little. Slow progress has been made to date. It is clear that, unless we address this in a meaningful way, there will be major problems for future generations.

Governments and others are now showing some concern over the lack of performance in reducing carbon. Little has been achieved, despite the considerable effort that has been aimed at the subject from many quarters. It is disappointing that, while the need to reduce carbon is obvious, in certain areas it is equally obvious that this is not yet a majority issue or one that has much current impact.

Many in the regulatory world see this lack of response as serious. Significant effort is now taking place to ensure performance in use matches design. In all areas of the world where carbon is taken seriously, performance in use is about to become a reported and regulated requirement. This will drive the clarity that is needed to ensure carbon issues are taken seriously. That will then open the door to sensible consideration of whole life carbon and, as long as everyone involved can respond, improvement will occur.

This will mean some changes to the way in which the team approaches a project, starting with ensuring the briefing is correct. This should be approached from the outset in the following way.

Consultant's brief and carbon specifics

Consultants need to pick up on the carbon agenda and address it. This should be possible, at least where there are reliable and consistent data. However, there may be extreme pressure to stay with the 'usual' agenda. Designing to achieve performance poses a real challenge, as there may be great disparity in the data that ensure the engineer, architect, etc. can react correctly to the requirements.

Legal requirements and the supporting regulations are increasing the pressure to ensure carbon issues are addressed, not just in design, but also in performance delivery.

Establishing what is needed is crucial, but difficult to define and encapsulate in a way that is practical and useful. When we have more reliable data, this will become more achievable. In Europe and the UK, we have a mix of regulations based largely on aspirational standards that are incomplete, and data that are somewhat questionable and limited. However, an increasing degree of practical experience is helping to improve this situation. It is hoped that these will mature so that the principles noted here can be used in a joined-up and productive manner.

In addition to the design, projects also depend heavily on the procurement and quality of construction to deliver what is needed. Looking to the construction sector, there needs to be a complete rethink in terms of the delivered product. Contractors need to approach their projects with a strategy.

Contractor requirements and brief

It is important to ensure specifications do not deviate from the required performance – not at all easy in our current cost-driven world. Quality and workmanship must drive the project on site, and the objectives much be clear to all involved. Contractors must be clear about what is required and why, crucial issues that are often lost or not considered.

Work on site must begin with a commitment to achieving success – many projects fail before they start. A positive approach is needed, with a drive to ensure that project requirements actually do reflect the finished, operational building that the client and the brief have defined.

Procurement needs to consider transport options, embedded carbon, and the relationship to the required performance. The options of distant manufacture versus material sourcing and local production must be assessed. The present industry is committed to the lowest cost above all else – this must give way to a carbon-costed answer, which in turn should lead to consideration of life attributes. Relative carbon values should be recorded for various procurement alternatives, just as energy performance options are commonplace currently.

How much does transport add to the carbon content of a material? What are the local manufacturing conditions contributing to the carbon embedded in the product? Not enough data are available to answer these questions reliably.

Issues from source to site include:

- sourcing of materials
- local manufacture infrastructure and environment
- local transportation
- transport to regional assembly or further manufacturing process
- transport to site.

Construction

The aim is to ensure the building is constructed efficiently to specification and with minimal waste. This should result in minimum carbon content of the materials and assemblies at the site. The new world of BIM should help here – nothing should be included on a project that has not been verified and checked for fit, performance, carbon content and quality.

Virtual reality augmentation is also just around the corner. This offers the possibility of a visual check on site of what is expected prior to, during, and on completion of construction. Ensuring what we have is precisely what was asked for is rarely achieved currently.

Completion

This stage involves checking the performance quality and fit to the user's requirements. Buildings have to be able to deliver performance while accommodating the users. Many currently fail due to the users following a pattern that was not considered or that is at odds with the designed methodology.

Handing over in a programmed sequence will always help at this point, but this approach tends to be used only on the very largest of projects, whereas it should be the norm. Bringing the performance of a building up to speed alongside the users becoming familiar with it is very desirable, and may help to achieve low-carbon objectives.

Maintenance

Checking the performance is optimum, getting the best out of every element, is crucial not only for carbon, but as a driver to achieve optimum life. At the top of the list for substantial change is maintenance, including checks on the building's systems, materials and wellbeing. All should be run at optimum. Cleaning is critical to ensure that materials age only as intended. Lack of cleaning will increase ageing. Repair and upkeep is part of this and must be part of day-to-day operations.

In summary, it is important that:

- the design is right
- procurement is clear and provides a precise description of what is needed
- the construction process is well managed
- handover is smooth
- occupation is a smooth transition to full use
- ongoing maintenance is focused, controlled and disciplined.

Feedback

Entering a new era for the built environment means that it is essential to check theory in practice and ensure continuous improvements are made. Taking this to the next level will involve using the momentum of the carbon-reduction programme to drive the need for whole life values to become a reality during the life of a building, instead of being a piece of theory used only in the early years or prior to construction.

Most of those who talk of whole life costing with passion are frustrated that this obviously logical way to treat the built environment is often put in the filing cabinet very early on in a project, never to reappear. The problem here is there has been no driver. Carbon reduction is the best thing to come along for whole life costing since its beginnings. There is a hope that where logic and good business have not driven better whole life consideration, regulation, or pressure from insurers, may achieve this. We are already seeing insurance companies being increasingly concerned over losses in the built environment, and maybe they will interest themselves in a least replacement-cost strategy that focuses on the real need for whole life costing in the real world.

Chapter 24

Final thoughts

This has been a fascinating journey. In many ways, thinking about whole life is a very enlightening experience. It is not often that we stand back and consider exactly how things come into being in order to provide a useful life and then disappear. We are often too interested in the effects they have, whether good or bad.

It is apparent, however, that if we spend more time looking from a measured, detached perspective, we might arrive at different and better answers. Often the drivers and relationships brought to bear are relatively superficial and not that important. A better, more challenging look at any enterprise will result in some surprising answers.

Often the response to a problem may be 'I wouldn't start from here'. Because of other pressures, we often approach problems from a position that will undoubtedly lead to a distorted conclusion. So it is with whole life questions about processes, about measurement, and about a whole range of other issues. Often is it clear that, while we start with plenty of good intentions, they prove to be neither robust nor consistent.

The plea of this book is for logic to prevail – so often, in practice, this is the hardest route to follow. If we can achieve this, it will be good for everyone – good for today, and good for tomorrow.

Index

Routledge
Taylor & Francis Group